FOREST FIRES

A Reference Handbook

Other Titles in ABC-CLIO's
CONTEMPORARY
WORLD ISSUES
Series

Books in the Contemporary World Issues series address vital issues in today's society such as genetic engineering, pollution, and biodiversity. Written by professional writers, scholars, and nonacademic experts, these books are authoritative, clearly written, up-to-date, and objective. They provide a good starting point for research by high school and college students, scholars, and general readers as well as by legislators, businesspeople, activists, and others.

Each book, carefully organized and easy to use, contains an overview of the subject, a detailed chronology, biographical sketches, facts and data and/or documents and other primary-source material, a directory of organizations and agencies, annotated lists of print and nonprint resources, and an index.

Readers of books in the Contemporary World Issues series will find the information they need in order to have a better understanding of the social, political, environmental, and economic issues facing the world today.

Library of Congress Cataloging-in-Publication Data
Omi, Philip N.
Forest fires : a reference handbook / Philip N. Omi.
    p. cm. — (Contemporary world issues)
  Includes bibliographical references and index.
    ISBN 1-85109-438-5 (hardback : alk. paper) — ISBN 1-85109-443-1 (ebook) 1. Forest fires. 2. Wildfires. 3. Fire management. I. Title. II. Series.
  SD421.O64 2005
  634.9'618—dc22

                                    2005003599

09 08 07 06 05 10 9 8 7 6 5 4 3 2 1

This book is also available on the World Wide Web as an eBook. Visit abc-clio.com for details.

ABC-CLIO, Inc.
130 Cremona Drive, P.O. Box 1911
Santa Barbara, California 93116-1911

This book is printed on acid-free paper.
Manufactured in the United States of America

FOREST FIRES

A Reference Handbook

Philip N. Omi

**CONTEMPORARY
WORLD ISSUES**

A B C CLIO

Santa Barbara, California • Denver, Colorado • Oxford, England

Contents

Preface and Acknowledgments

The purpose of this book is to explore the many dimensions of forest fires and their impacts, both in North America and elsewhere. Readers will develop an appreciation for current practices as well as for historical precedents. Information will be presented on the contributions of pioneer leaders and current practitioners in the study and management of fires.

An understanding of forest fires requires an inquisitive mind, though clear-cut answers to questions are not always possible: fires are extremely complex phenomena and are often tinged by human experiences. Although murky, the crystal ball for the future will be discussed when speculation is possible. Pathways to further study in print and electronic media are identified, along with fire management agencies that represent important information resources.

Forest fires do not respect geopolitical boundaries but affect societies and land management practices worldwide. Basically, a fire will burn anywhere that plants and aboveground biomass will support combustion. A fire alters the distribution of ecosystems, as well as their structure and processes. Readers will learn of some of these consequences—although the major focus will be on U.S. ecosystems and fire practices, especially by agencies with fire management responsibilities. Even so, fire effects and management practices may be of interest to readers from other countries hoping to gain insight into the evolution of U.S. responses to its fire problems.

This book is aimed primarily at those who seek an overview reference to fire science and management, with focus on highlights yet without many of the extra details included in textbooks. Thus public library users and journalists, as well as high school and first-year college students, will find insight into problems and challenges, chronologies, key players, facts and data, and print and nonprint resources. Those seeking greater detail will learn of primary source materials and websites that should guide more detailed learning opportunities.

Seven chapters and a glossary are included, as per the Contemporary World Issues series format. This format provides readers with a good starting point for understanding the scope of forest fire influences on our lives.

Chapter 1 (Forest Fires: In Context) provides a framework for thinking about forest fires and their impacts in both traditional and nontraditional ways. Traditionally, forest fires have been considered as universally destructive and undesirable—an understandable perspective given our preconceived ideas about fire and news media depictions of blackened landscapes. Every fire season we are bombarded with graphic images of wildfires that strike homes and fires that invade parks and wildernesses. Yet fires can be friendly; in fact, fires burned with less damaging consequences prior to Euro-American settlement. Humans have complicated matters by building their homes, grazing their animals, suppressing fires, and allowing fuels to accumulate as a consequence. We are beginning to realize that destructive fires may partially be the result of our own mismanagement of the forest. Restoring forests to presettlement conditions may provide a solution for some ecosystems, but might not be possible or desirable everywhere. Still, perhaps wildfires would not seem so destructive if our forests were less choked with fuels—as in those halcyon early days. Implications for fires of the future are also discussed.

Chapter 2 (Problems, Controversies, Solutions) focuses on the predominant traditional and nontraditional views toward fires. The traditional fires presented in Chapter 1 are given greater scrutiny by probing into characteristic issues and questions. For example, who is responsible when wildfires strike homes? Why do killer fires continue to occur despite lessons learned from the past? Can we allow fires to burn in park and wilderness areas that, after all, were partially established to provide living laboratories for us to learn about natural processes? Is prescribed fire a viable tool for managing landscapes? These and other probing questions are examined from a variety of perspectives.

The history of North America is punctuated by fire episodes. In Chapter 3 (Chronology) we examine a time stream of historical fires, including Native American firings that were practiced before European settlement of the continent. Historical fires are classified according to their predominant effects, including on homes, on humans, and on natural resources. Arranging historical fires by discrete time periods in U.S. history provides insights into the evolution of thinking about fires and forests.

The study and management of forest fires has attracted many gifted thinkers and practitioners, some of the most noteworthy of whom are described in Chapter 4 (People and Events). Linkage of historical time frames with legislation and events provides us with additional insights into our responses (as a society) to fires.

Chapter 5 (Facts and Data) presents trends and overviews as to the status of our knowledge about wildland fires and their many effects. Management strategies and tactics are also discussed, as well as brief insights into fire management costs. Much of the material in this chapter is elaborated in greater detail in formal textbooks on fire. Here, I have attempted to condense our understanding into a single chapter, focusing on the highlights of each topic.

The different perspectives on fire presented in Chapters 1 and 2 have influenced a wide range of public and private agencies, organizations, and interest groups with a stake in better understanding the role of fire in forests. Chapter 6 (Directory of Agencies and Organizations) presents the myriad of entities (including contact information) that inform our thoughts about forest fires.

Chapter 7 (Print and Nonprint Resources) summarizes different sources for information about forest fires, including websites. Here again, I have attempted to condense a significant proportion of the countless sources that are available on this subject. I make no claim to have captured every significant resource, but the interested reader will be given numerous avenues for further study.

The Glossary is intended to assist readers with some of the terminology and jargon that is part of every professional and technical field. In it I have picked and chosen definitions from a variety of sources, as well as interjecting a few of my own.

The limited use of graphics or figures and tables is in conformance with ABC-CLIO editorial policy for reference books. For the study and understanding of fire, this may seem an unreasonable restriction. Nonetheless, I have done my best to conform to the publisher's policy.

I wish to thank Jeremiah Sisneros for assistance with graphics, and biographical information, and Annie Brown for assistance with biographical information. Also I wish to thank the Joint Fire Science Program, USDA Forest Service Southern Forest Experiment Station, McIntire-Stennis Cooperative Forestry Research Program for support on research projects that provided results of use to this reference book; and ABC-CLIO for supporting this effort. In particular, I wish to thank Alicia Merritt for help during draft stages, Martin Hanft for helpful editorial suggestions, and Carol Smith and Cisca Schreefel for guidance during the revision process.

Last but not least, I thank Nik Omi and Mei Nakano, for patience and understanding throughout this whole project.

Common Conversion Factors for Use in Interpreting Data

Length
1 cm   0.39 in
1 m   3.3 ft
1 km   0.62 mi

Area
1 ha   0.405 ac; 10,000 m^2 = 2.47 ha
1 km^2 0.39 mi^2 = 247 ha

Liquid Volume
1 liter   0.265 gal

Temperature
Let t = temperature (°C)
Temperature (°F) = t (9/5) + 32

Heat and Heat Content
1 cal   4.18 joules = 0.0039 Btu
1,000 Btu/lb   2.324 MJ/kg

Fuel Loading
1 metric ton (tonne)/ha 0.45 ton /ac

Power/Irradiance
1 kw/m^2 5.3 Btu/ft^2-min

Fireline Intensity
1 kw/m   0.29 Btu/ft-sec = 0.238 kcal/m-sec

1

Forest Fires: In Context

This chapter provides an overview and introduction to forest fires. Fire has been a part of most terrestrial ecosystems for millions of years. Prehistoric humans used fire extensively, and societies later harnessed fire to fuel the Industrial Revolution. So we know that fire can be of tremendous benefit to humankind if properly used, even in forested landscapes. Yet fires in the forest often conflict with human values. To develop a better appreciation for some of those conflicts, we examine case studies of different types of fires that cause management dilemmas. We conclude this chapter by looking at fires in other parts of the world and considering a prognosis for the future.

To most people, forest fires conjure up images of blackened destruction, including dead trees, displaced wildlife, and devastating loss of human life and property. Indeed, forest fires sometimes produce tragic and costly outcomes, and in the short term, a burned area may not be visually pleasing. But a typical forest fire produces a mosaic of fire effects, including some areas that burn with low intensity with relatively little damage to the tree canopy. Several of the many faces of fire include: benign to active surface fires and raging crown fires; fire effects that range from little or no effect to a gallery of dead trees (or snags); and postfire mosaics that vary from large, uniform burn scars to small, irregular patches. The well-intentioned Smokey Bear campaign of the U.S. Forest Service has reinforced the negative stereotypes associated with fire by advocating all-out elimination of fire in the forest. Fire suppression is surely required where human developments are threatened. But to focus solely on fearful outcomes reveals only part of the story about wildland fires. In fact, forests have burned for millennia on earth, and natural ignition sources (such as lightning, volcanoes, or spontaneously ignited coal seams) play an important role in the structure and function of natural ecosystems. Many of the trees, shrubs, and herbaceous vegetation in areas subjected to repeated fires develop adaptations that make possible survival and persistence despite periodic burning. In fact, some plants and animals thrive in fire-prone environments.

Fires in the forest play an important role in decomposing organic biomass and cycling nutrients. Without fires, insects, and fungi that decompose plant and animal material, most forests would be choked with detritus and dead materials, largely impenetrable to wildlife and humans. Combustion of forest fuels speeds the essential process of decomposition that is always ongoing in wildland areas by oxidizing flammable biomass into carbon dioxide, water, and numerous other chemical compounds—some of which are essential for plant growth, others of which may be detrimental to humans and other life forms. In a forest fire, dead and live vegetation is converted to heat, light, and sound, often in spectacular fashion.

Thus the story of forest fires is much more complex than that usually conveyed in newspapers and television sound bites every summer in the western United States. Fires receive a lot of press coverage because they present spectacular photo opportunities and because natural environments are cherished by society for aesthetic and economic reasons. And fires touch us because they affect human lives—not only home owners and recreationists in fire-prone environments, but also the land managers charged with stewardship of those areas and the fire crews who take their orders from the managers. Furthermore, researchers from a variety of life, physical, and social sciences spend their entire professional careers studying fire and its effects.

Major themes in the study of fire include fire behavior, fire ecology, and fire management. These fields have spawned a myriad of subdisciplines that are simultaneously distinct and intertwined with one another. For example, fire behavior cannot be understood without a thorough understanding of wildland combustion and fire meteorology. Understanding fire ecology requires knowledge of the historical role of fire in ecosystem dynamics, as well heat effects on specific organisms. Fire management encompasses activities required to control a fire's spread as well as preparedness and mitigation strategies, which might include the intentional use of fire to achieve land management objectives. After a wildfire is successfully suppressed, focus shifts to rehabilitation efforts: revegetating denuded hillsides, mitigating erosion on- and off-site, and reducing scars caused by fire control activities.

From a broader perspective, fire can be considered among several disturbances (for example, floods, grazing, windstorms, insect epidemics) that affect landscapes and natural ecosystems. Many plant communities are dependent on disturbances, especially for regeneration (Hobbs and Huenneke 1992), so land managers must take these disruptions into consideration in their plans for an area.

Throughout this book I will often interchange the term "forest fires" with more contemporary allusions to wildland fires. In reality, a forest is a complex assemblage of vegetation types, such as trees, grasses, shrubs, mosses, and lichens—all of which may burn under the appropriate mix of temperature, wind, and moisture. The mosaic of flammable vegetation types, commonly referred to as wildland in the United States and elsewhere, may be known by other terms, such as bush (for example, in Australia and Canada) or *bosch*/veld (South Africa). When wildlands burn, people most commonly associate such fires with forest, perhaps out of long-standing tradition and perceptions of dominant vegetation (that is, trees) aflame. To separate out burning forest from other types of vegetation would be cumbersome and overly restrictive for most purposes. For example, most wildfire statistics do not differentiate between vegetation types, either with respect to point of origin or area burned.

The purpose of this book is to present the many dimensions of forest fires and their effects, not only in North America but elsewhere as well. Readers will develop a sense for historical events and precedents to current practices, as well as learning of pioneer leaders and practitioners in the study and management of fires. Pathways to further study in print and electronic media are identified along with fire management agencies. The book concludes with a glossary of common terms in fire science and management.

U.S. Fire Activity

Every year in the United States, hundreds of thousands of fires burn in forests, shrubfields, and grasslands. Even peat bogs in the upper Great Lakes states or swamplands in the Southeast will carry fire following a prolonged drought. Typically, most fires burn and go out with little fanfare because of their limited extent or impact, whereas relatively few (that is, fewer than 3 percent of the total ignitions) may cause most of the burned area, costs, and losses. Major fire activity is episodic over time, with some years causing more damage and concern than others (see Figure 1–1). For example, the year 2000 produced an extraordinary fire season by most standards. More than 122,000 fires burned over 8.4 million acres (3.4 million hectares) in the United States during the year, destroying or damaging more than 850 homes and costing over U.S.$2 billion. A greater number of fires in 1995 (130,000) produced only 27 percent as much burned area as the year 2000; thus in most years the number of ignitions is generally less important than the area burned. In fact, as a society we generally pay little attention to most fires, tending to focus on the larger, more spectacular ones.

Interestingly, most of the areas burned in major fire years, such as 2000 and 2002, have burned and reburned for millennia, through an endless cycle of biomass removal by fire, forest regeneration and growth, and subsequent ignition. Lightning provided the primary ignition source for these historical fires, at least prior to the time of the aboriginal humans, although other ignitions were started by volcanoes or spontaneous combustion (for example, in subterranean coal seams). Prior to the arrival of Europeans in North America, aboriginal burning was common. Native Americans burned vegetation for numerous reasons, including herding of wildlife, warfare, and communications.

As colonists pushed westward in North America, they encountered Native American firing practices and also found fires useful for a variety of settlement purposes, such as clearing land. Thus human-caused ignitions more recently have augmented the natural causes for forest fires. Some of these fires have left indelible imprints because of their size, loss of human life and property, or natural resource impacts. Some of the most notable fires are noted in Table 3–1 in Chapter 3 (Chronology).

Historical fires also have influenced the evolution of forestry and natural resource policy, as laws and ordinances were passed, sometimes in reaction to mounting losses caused by fire (Chapter 4). Concurrently, agencies charged with managing fires on public and private lands were spawned and they developed policies to contend with perceived threats to society from wildland fire.

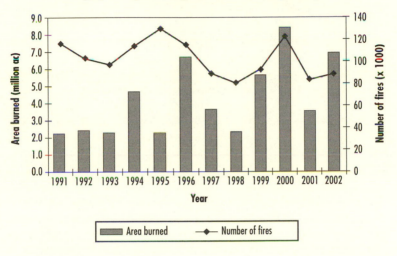

Figure 1-1
U.S. Fire Frequency and Area Burned, 1991–2002
The number of fires and area burned will fluctuate from year to year and decade to decade, depending on climatic variations (e.g., prolonged drought) and fuel susceptibility.

Source: National Interagency Fire Center 2002.

Thus the story of forest fires also involves people, those affected by wildland combustion as well as those who attempt to manage the effects of fires.

The fires that garner the most attention are those that affect human life and property, or natural resources of value to society (such as wildlife, timber, range, or water). Forest fire scientists also study fires that may not be so spectacular or have little apparent impact, thus acknowledging that fires can affect a full range of human values. Although every fire is unique in its behavior and effects, we will start by describing case studies of fires that have generated the most interest (and controversy) in the United States. As individual incidents and as a group, these fires exemplify society's fears and hopes (for eventual management) concerning forest fires, including both incidents that are considered highly undesirable and the lesser-known fires that illustrate desirable effects. Undesirable fires destroy homes, kill firefighters, and harm or destroy natural resources. Desirable fires reduce fuel hazards or are otherwise beneficial in terms of ecosystem structure and function. Not all fires fit uniformly into either category (that is, desirable versus undesirable). Some fires may result in a mosaic of desired and undesired results over a landscape. Desirable attributes may be found in even the most destructive fires (for example, creation of desirable wildlife habitat for some species). Also, some desired fires can be transformed into ugly holocausts if fanned by extreme winds or if convective plumes dominate the fire spread.

Wildfires that Strike Home

Perhaps the most spectacular (and tragic) fire is one that directly destroys human life and valuable property, particularly houses. The Tunnel fire of 1991 in the east Bay Area (Oakland and Berkeley) of California stands out as a classic example of a wildland fire that burns homes. Many residents were unaware of the wildfire risks in this area, despite historical precedents for fires starting in flammable vegetation and spreading to nearby homes—such as the 1923 Berkeley Hills fire that destroyed 584 homes north of the University of California campus (Chapter 3). As a consequence of their ignorance, most residents were unprepared for the events of October 20, 1991.

The Tunnel fire started innocuously enough, with a small brushfire in the East Oakland hills that was quickly attacked by Oakland fire department employees on Saturday, October 20. Firefighters from the Oakland fire department attacked the burning brush for about two hours and "mopped up" glowing embers until dark, before leaving the area. The next morning, crews returned to the fire to continue mop-up and retrieve hoses left behind the day before, only to encounter high winds that had caused the fire to flare up and escape control. By nightfall Sunday, more than 1,600 acres (648 ha) of expensive real estate was in ruins, destroying 3,354 homes and 456 apartments, leaving 5,000 residents homeless, injuring 150 people, and leaving twenty-five dead (Adler 1992). The fire burned along four major fronts within an area of 5.25 square miles, resulting in more than $1 billion in fire damages (Ewell 1994).

The 1991 Tunnel fire is a classic example of a fire that starts in a wildland area and then burns into an urban area. The southern California fire complex of October 2003 provides another vivid example, in which ten fires in five counties burned almost

750,000 ac (304,000 ha), destroyed more than 3,650 structures, and left 22 dead. The subsequent Christmas 2003 flood left fourteen dead and several missing as intense rain storms pelted the burn areas, causing massive sediment flows in Lytle Creek Canyon. Other examples are noted in Chapter 3 (Chronology). Fires in the so-called urban interface (also known as the urban-wildland interface or exurban areas) present increasingly difficult challenges for fire managers, especially as more people choose to live in formerly wild areas in pursuit of the American dream (for example, a home in the woods).

Killer Fires

All the fatalities in the 1991 Tunnel fire were residents in the East Oakland hills of the San Francisco Bay Area. Occasionally casualties include firefighters attempting to control a fire. Firefighting is hard work, yet certain individuals are drawn to this risky occupation. Whether for the excitement, the arduous challenge, or the lure of wages inflated by overtime pay and hazardous duty, as many as 20,000 women and men nationwide may join the firelines annually. Both permanent and seasonal employees (primarily college students) recognize and accept the dangers inherent in fighting fires, yet fatalities occur with some regularity—partly on account of failure to anticipate the awesome power and destruction that a wildland fire can display. The usual causal factors, including shifting winds, low moisture content of plants and dead fuels, and firefighter fatigue or misjudgment, are well known and discussed in firefighting training courses. Furthermore, firefighters are provided with guidelines for building a fire control line safely—such as posting of lookouts, maintenance of communications, establishment of escape routes, and creation of safety zones, after Gleason (1991). Yet a fire can blow up with little warning and trap those who are unwary of the inherent dangers or who miss the cues provided by the fire and its environment.

The 1994 South Canyon fire provides a classic case study of fires that result in firefighter fatalities. The fire was started by lightning on July 2 west of Glenwood Springs, Colorado. Initially the fire attracted little attention, because thirty or forty other fires had been ignited that day by lightning throughout the state (Junger 2001); furthermore, the fire was burning in an area with relatively low resource values, safely away from any residences.

By noon on July 4 the fire size was approximately 3 acres in size, spreading downslope in leaves, twigs, and dead grasses. By the morning of July 6, the fire covered 127 acres and was considered a growing threat. Elite smokejumpers and hot-shots were sent to assist a Bureau of Land Management crew constructing firelines along several flanks of the fire. Late in the afternoon high winds (30 to 45 mph) hit the area and caused several uphill fire runs. As the fire spread north and east, it formed a high-intensity, fast-moving, continuous front in the live green Gambel oak (*Quercus gambellii*) canopy (Butler et al. 1998). As the fire moved uphill, it overran firefighters attempting to climb to safety along the fire-line. Fourteen firefighters perished when they were unable to out-race the wall of flames moving up the hillside.

Tragically, the fatalities occurred despite lessons provided by previous tragedy fires, most notably the 1949 Mann Gulch fire in which eleven smokejumpers perished under somewhat similar circumstances. In 1976 four firefighters lost their lives in similar vegetation and topographic conditions. In fact, fatalities persist despite knowledge and experience. Firefighter fatalities cannot be justified under any circumstances, no matter how valuable the natural resources or homes being protected. In Chapter 3 we examine numerous other fires that resulted in firefighter death and injury, and look at common denominators among fatality fires.

Park and Wilderness Fires

Many forest fires don't kill people or damage homes but nonetheless are costly and of great public concern. The South Canyon fire was largely ignored in its initial phases because houses were not considered at risk; numerous other fires were assigned higher priority for suppression action. Fires that burn in parks and remote wilderness areas similarly may not threaten homes, but they are still important to society because people like to visit wild and historic areas. Natural and human-caused ignitions have played an important role in the evolution of these remote and wild ecosystems for thousands of years. Such areas provide important laboratories for studying fires and their effects, providing glimpses into the historical role of fire in wildland ecosystems, including plant communities that have burned recurrently over hundreds of thousands of years. Furthermore, attempts at excluding fires from these areas may be undesirable, since much of the vegetation evolved with recurrent fire; studies have shown increased fuel levels associated with fire exclusion. At the same time, fire protection must be provided to park developments such as visitor centers, administrative headquarters, and park concessions (such as restaurants and hotels). Thus fire in parks and wildernesses presents special management challenges and dilemmas. The challenges include balancing the desire to restore fire as a natural process in wildland ecosystems while protecting life and property—say, in a national park. This dilemma was crystallized in 1988, when Yellowstone National Park experienced an unusual fire season—one generating more national and international attention than any other event in the history of the area and involving the largest firefighting operation in U.S. history (Greater Yellowstone Coordinating Committee 1989). More than half a million hectares (actually about 1.4 million acres) burned with differential severities and effects during a four-month period (June–September) in and around the "crown jewel" of U.S. national parks.

Starting in the middle of June 1988, a series of fires were ignited by lightning and other causes in the Greater Yellowstone Area (consisting of Yellowstone and Grand Teton national parks, and adjacent national forests). Some of the lightning fires were allowed to burn, so as to restore fire as a natural process in predetermined low-threat areas, as stipulated in Yellowstone's fire management plan. Furthermore, scientists have speculated for years that natural fires might become self-regulating as burned areas create buffers that would regulate the size and severity of future fires. In fact, 235 lightning ignitions had been allowed to burn an average 240 acres each since Yellowstone National Park had implemented its natural fire policy in the early 1970s (Simpson 1989). Apparently, eleven of twenty fires that had been allowed to burn in June 1988 under the auspices of the park's natural fire policy had gone out on their own (Williams 1989), despite coming on the heels of the driest spring on record. Unfortunately, the moisture situation worsened considerably when expected June and July rains failed to materialize. By mid-July, Park fire managers were confronted with more fire than desired on the landscape; by the end of July, the natural fire policy for the park had been rescinded, and all fires were being managed under full suppression guidelines. Unfortunately, by that time the stage had been set for one of the most calamitous and expensive fire campaigns in U.S. history.

Some of the Yellowstone fires were allowed to burn in order to allow a more natural role for lightning fires in the park. Others (human-ignited fires) were fought as wildfires from their inception. Our next case study looks at fires that are intentionally set to achieve management objectives.

Friendly Fires

Researchers have known for years that low-intensity surface fires have regularly burned through low-elevation ponderosa pine (*Pinus ponderosa*, Laws.) forests of the western United States (Arno and Brown 1991; Agee 1993; Covington and Moore 1994; Mutch 1994; et al.). Historically, these fires were caused by lightning and the firing practices of Native Americans. However, twentieth-century fire suppression efforts restricted or eliminated many naturally ignited fires that might have grown larger and covered wider areas of the forest. As a consequence, fuels have accumulated in the areas from which fire has been systematically excluded, leading eventually to larger, more severe wildfires once the inevitable ignition occurs. In essence, systematic fire exclusion led to replacement of frequent, low-intensity fires (that is, friendly fires) with infrequent, stand-replacing fires (Biswell 1989). Thus, the notion of prescribed fire was developed to describe fires intentionally ignited to achieve land management objectives, such as fuel hazard reduction or habitat improvement for wildlife. Actually, the notion that fire could be useful has been around for centuries, perhaps best embodied by our understanding that native peoples used fire extensively as a tool for rejuvenating grasslands used by favored wildlife species, such as bison, or for other purposes, such as maintaining travel corridors or signaling (see Chapter 3). As early as the 1930s, foresters in the southeastern United States recognized the potential usefulness of periodic fire for managing longleaf pine (*Pinus palustris*) stands or for managing wildlife habitat (Pyne 1982).

Public agencies with fire management responsibilities in the United States have recognized the potential benefits and cost-effectiveness of prescribed fire since the 1970s. The National Park Service was among the first federal agencies to recognize the natural historical role of fire in ecosystems as diverse as the Florida Everglades (Bancroft 1976) or the Giant Sequoia (*Sequoiadendron gigantea*) groves of the southern Sierra Nevada (Hartesveldt et al.

1975; Kilgore and Taylor 1979). Systematic fire exclusion as practiced for most of the twentieth century had led to excessive fuel accumulations or changes in ecosystem structure and function. Rather than attempting to exclude fire from these systems by systematic fire suppression, park managers thus began experimenting with intentional firing operations under carefully prescribed fuel, weather, and topographic conditions to achieve prespecified management objectives.

As useful a tool as fire may be in manipulating vegetation, it sometimes may have unintended consequences. The escaped Cerro Grande prescribed fire provides a classic example of a prescribed fire run amok. On May 4, 2000, fire management personnel in Bandelier National Monument near Los Alamos, New Mexico, ignited a prescribed fire in the late evening. The objective was to reduce fuel hazards in one corner of the National Monument, which was bordered by adjacent private land and national forest. Sporadic winds on May 5 caused embers to cross control lines, and the fire was declared a wildfire. The wildfire was contained on May 6, but by May 7 winds again breached control lines and caused the fire to escape to the east into the Santa Fe National Forest. By May 10, the towns of Los Alamos and White Rock were in the path of a raging wildfire that eventually destroyed 235 homes and numerous other buildings. The wildfire was not declared controlled until July, by which time it had incinerated 47,640 acres (19,287 hectares) and cost more than $1 billion in federal expenditures (Los Alamos National Laboratory 2000–2001).

Ironically, much of the area that burned in the Cerro Grande fire within the Santa Fe National Forest included vast areas of ponderosa pine, which had historically burned frequently with low-severity surface fires. Researchers have long known that mature ponderosa pine trees have thick bark and characteristically self-prune the lower branches, thereby enabling them to survive low-intensity surface fires. Forest managers had initiated preventive fuel treatments (such as prescribed fire, mechanical thinning, or both) in strategic areas prior to the occurrence of Cerro Grande wildfire, in order to reduce the severity of future wildfires in the area. In fact, mechanical thinning followed by prescribed fire reduced ponderosa pine tree mortality when compared with adjacent, untreated areas (Omi and Martinson 2002). Those treatments had been implemented as part of a broader ecological restoration plan to replicate forest conditions that had been prevalent prior to the arrival of European settlers in the Southwest. At that early time, the combination of periodic lightning ignitions and native firings removed fuels in the understory, including susceptible vegetation and younger, fire-intolerant saplings. Fire-exclusionary practices after the turn of the twentieth century, coupled with Indian displacement to reservations, essentially removed ignition sources, resulting in a much more dense and flammable forest than in the past. Today, fires that burn through present-day ponderosa pine areas often burn as stand-replacing crown fires. Thus, managers in the Santa Fe National Forest had hoped eventually to restore presettlement forest conditions, including a more open, parklike forest structure predominated by larger trees with fewer younger trees—a condition that would once again tolerate periodic surface fires. Unfortunately, the Cerro Grande fire burned through before those plans could be implemented, once again suggesting that our forests cannot be sustained by fire exclusionary practices. Even so, the Cerro Grande fire exhibited typical variability as it spread across the landscape, including pockets of lightly burned areas within the fire mosaic.

Other Classic Forest Fires

The fires noted above are not the only incidents that captivate our imagination. For a listing of the more significant incidents in U.S. history, see the chronology in Chapter 3. Perhaps the most storied forest fires in U.S. history occurred early in the twentieth century. The great 1910 Idaho and Montana fires burned more than 3 million acres, mostly during a single two-day period (August 20–21), during which smaller fires coalesced under the influence of winds and the burn area tripled in size. More important, the 1910 fires severely tested the fortitude of the fledgling U.S. Forest Service, fueled debates about pending timber famine, and set the stage for a concerted national effort to exclude fire from natural ecosystems that would endure for the entire twentieth century (Pyne 2001). This campaign of fire exclusion persists today, despite ecological evidence suggesting the futility of such efforts and the spiraling costs of suppression.

Other notable fires that generated considerable interest during the twentieth century include the 1932 Tillamook fire in western Oregon, the 1956 Bel Air fire in southern California, and the Mack Lake prescribed fire escape in 1971. The Tillamook fire burned 236,000 acres of prime timbered land, including more than 200,000 acres in one major run occurring on August 25, 1933 (Morris 1935). The Bel Air fire signified the incursion of fire from the forest into modern suburbia. The Mack Lake fire, set initially to improve the habitat for the endangered Kirtland's warbler, escaped control lines, burned up forty-four homes, killed one person working on the fire (Simard et al. 1981), and almost ended the prescribed burning program of the U.S. Forest Service.

Forest fires occur elsewhere on the planet. Notable recent fire episodes occurred in South Africa (2000), Ethiopia (2000), Australia (2001 and 1994), Malaysia (1999), Mexico (1998), and Indonesia (1997–1998). Most areas burn annually, such as the Mediterranean region in Europe, the tropical forests in South America, sub-Saharan Africa, and southeastern Asia. Annual savanna burnings on the African and Australian continents may be the largest of all burns, yet they may be the least understood in terms of environmental effect. Fire impacts and societal responses vary by locale, though fires will burn wherever fuels, weather, and topography will permit. Generally, the occurrence of forest fires is viewed negatively by most nations, although scientific evidence in places such as North America and Australia (collected from tree rings, charcoal in lake sediments, and aboriginal oral histories) supports the pervasive history of fire going back many years.

Fires of the Future

As evidence mounts for the likelihood that human activities are partially responsible for global warming, concerns will continue to grow regarding the role of fire in a warmer environment enriched with greenhouse gases such as carbon dioxide (CO_2), methane (CH_4), and nitrogen dioxide (NO_2). Each of these gases is a product of forest combustion, although CO_2 is produced in much greater quantities per ton of fuel consumed. From a global perspective, though, much greater attention is placed on the outputs of methane and nitrogen dioxide, especially by industrialized countries (for example, in the Kyoto Protocol). Despite the considerable controversies over possible consequences and the best approaches for managing global warming, the consensus belief is that the higher temperatures will be conducive to wildland fires. At the same time, to the best of our knowledge, global warming will contribute to heavier precipitation in some areas as well as prolonged drought elsewhere. Increased levels of CO_2

with warming and moisture could also accelerate plant growth and subsequent moisture levels. So fire activity could actually decrease in some areas but increase in others. Where the number of fires increases, firefighting infrastructures could be strained and smoke pollution might become more severe.

Even without global warming, the fires cited in this chapter provide perspective on the future of fire in the United States and elsewhere. Suffice to say that fires will continue to burn wherever fuel, weather, and topography permit. No amount of human intervention will succeed in permanently excluding fire from the forest, so the more realistic challenge is for humans to learn to coexist with fire, rather than to attempt to eliminate it. That will require a major shift in thinking on the part of forest managers, local officials, and people who live in or use fire-prone environments. Forest managers are beginning to recognize that the fire-exclusionary posture is not sustainable in the long term, and they are beginning to implement fuel treatment and ecological restoration programs aimed at reducing fire hazards while facilitating a more natural role for fire in the forest. Local officials (including those who regulate planning and zoning in forested areas) need to recognize the high fire risks associated with suburban or exurban development in fire-prone environments and manage population growth in those areas accordingly.

The biggest challenge will be to manage the people who live in and use the fire-prone forest. Humans sometimes appear remarkably naive when it comes to living with fire in the forest. Buildings are constructed with reckless abandon in vegetation types that have repeatedly burned in the past. Forest dwellers demand subsidies by the government (in the form of fire protection) for their individual decisions to build in a fire environment, yet they resist governmental efforts to legislate or mandate fire-safe buildings or defensible spaces. Ultimately, people must realize the risks of living with fire in the forest and bear the consequences if they choose to ignore the warning signals. We will return to the theme of people living in fire-prone environments in Chapter 2.

Literature Cited

Adler, P., ed. 1992. *Fire in the Hills: A Collective Remembrance*. Berkeley: Patricia Adler, Publisher.

Agee, J. K. 1993. *Fire Ecology of Pacific Northwest Forests*. Washington, DC: Island Press, 493 p.

Arno, S. F., and J. K. Brown. 1991. "Overcoming the Paradox in Managing Wildland Fire." *Western Wildlands* 17(1): 40–46.

Bancroft, L. 1976. "Natural Fire in the Everglades." Pp. 47–60 in *Proceedings Fire by Prescription Symposium*. USDA For. Serv. Southern Region, Atlanta, GA. Washington, DC: U.S. Government Printing Office, 127 p.

Biswell, H. H. 1989. *Prescribed Burning in California Wildlands Vegetation Management*. Berkeley: University of California Press, 255 p.

Butler, B. W., R. A. Bartlette, L. S. Bradshaw, J. D. Cohen, P. L. Andrews, T. Putnam, and R. J. Mangan. 1998. *Fire Behavior associated with the 1994 South Canyon Fire on Storm King Mountain, Colorado*. Research Paper RMRS-RP-9. Ogden, UT: USDA Forest Service, Intermountain and Range Experiment Station.

Covington, W. W., and M. M. Moore. 1994. "Southwestern Ponderosa Pine Forest Structure: Changes since Euro-American Settlement." *Journal of Forestry* 92(1): 39–47.

Ewell, P. L. 1994. "The Oakland-Berkeley Hills Fire of 1991." In *The Biswell Symposium: Fire Issues and Solutions in Urban Interface and Wildland Ecosystems*, February 15–17. Walnut Creek, CA: D. R. Weise and R. E. Martin (technical coordinators). General Technical Report PSW-GTR-158; Albany, CA: Pacific Southwest Research Station, Forest Service, U.S. Department of Agriculture, 199 p.

Gleason, P. 1991. "LCES—A Key to Safety in the Wildland Fire Environment." *Fire Management Notes* 52(4): 9.

Greater Yellowstone Coordinating Committee. 1989. "The Greater Yellowstone Postfire Assessment." J. D. Varley, S. E. Coleman, and B. C. Kulesza, cochairs. S. M. Mills (ed). Prepared by Beaverhead, Gallatin, Custer, Shoshone, Bridger-Teton, and Targhee national forests, and Grand Teton and Yellowstone national parks. Unpublished document.

Hartesveldt, R. J., H. T. Harvey, H. S. Shellhammer, and R. E. Stecker. 1975. *The Giant Sequoia of the Sierra Nevada*. Washington, DC: USDI, National Park Service, 180 p.

Hobbs, R. J., and L. F. Huenneke. 1992. "Disturbance, Diversity, and Invasion: Implications for Conservation." *Conservation Biology* 6(3): 324–337.

Junger, S. 2001. *Fire*. New York: W. Norton and Co., 224 p.

Kilgore, B. M., and D. Taylor. 1979. "Fire History of a Sequoia-Mixed Conifer Forest." *Ecology* 60: 129–142.

Los Alamos National Laboratory. 2000–2001. *Cerro Grande: Canyons of Fire, Spirit of Community*. Los Alamos, NM: Los Alamos National Bank.

Morris, W. G. 1935. "The Details of the Tillamook Fire from Its Origin to the Salvage of the Killed Timber." USDA For. Serv. Pacific Northwest Forest Experiment Station, Portland, OR, 23 p. plus app.

Mutch, R. W. 1994. "Fighting Fire with Prescribed Fire: A Return to Ecosystem Health." *Journal of Forestry* 92(11): 31–33.

National Interagency Fire Center. 2002. www.nifc.gov/stats.

Omi, P. N., and E. J. Martinson. 2002. "Effects of Fuels Treatment on Wildfire Severity." Final Report to Joint Fire Science Governing Board, March 25, 2002. 40 p. (also accessible at www.cnr.colostate.edu/~fuel/westfire /FinalReport.pdf).

Pyne, S. J. 1982. *Fire in America.* Princeton: Princeton University Press, 654 p.

———. 2001. *Year of the Fires: The Story of the Great Fires of 1910.* New York: Viking, Penguin Group, 322 p.

Ryan, K. C. 2001. "Fire Severity: Concepts and Applications." CD-ROM version of presentation to Colorado State University, Department of Forest Sciences, May 2001.

Simard, A. J., D. A. Haines, R. W. Blank, and J. S. Frost. 1981. "The Mack Lake Fire." USDA Forest Service General Technical Report NC-83, 36 p.

Simpson, R. W. 1989. "The Fires of '88. Yellowstone Park and Montana in Flames." Helena: American Geographic Publishing.

Williams, T. 1989. "The Incineration of Yellowstone." *Audubon* 91(1): 38–85.

2

Problems, Controversies, Solutions

Wildfires pose serious problems worldwide, though timing and circumstances vary. Episodic wildfires occur on every continent (except Antarctica), causing loss of life and property as well as disrupting normal commercial activities. Wildfire episodes recur in locales as varied, for example, as Siberia, southeast Asia, China, Ethiopia, South Africa, Australia, the Mediterranean, and North and South America, costing the equivalent of billions of dollars annually. Untold social displacement and psychological damage commonly occur when incidents clash with local values.

Uncontrolled firing of the forest has also been implicated as an agent of deforestation in the tropics, a global concern from the standpoint of the planet's biological diversity and possible loss of beneficial plant species as yet undiscovered. Often the tropical forest is cleared by itinerant peasant farmers who set fires to clear the land of unused tree stems, logs, branches, and leaves (otherwise known as slash—that is, taking its name from the cutting activity). The pattern of shifting cultivation, or "slash and burn" agriculture associated with migrant farmers attempting to subsist off the land, annually results in innumerable uncontrolled fires, with thick smoke blanketing entire regions. The cumulative effect of these practices is considered a significant social problem for sustaining tropical forests.

Wildfire effects occur on-site (where the fire is burning or has burned recently) as well as off-site (that is, downstream or downwind). Sometimes the downstream effects are delayed until the

17

onset of rainstorms. In steep terrain, rainstorms may trigger land-slides after a fire removes plant cover and kills root systems that formerly held the soil in place. Mudflows from unstable hillsides move downhill, knocking houses off their foundations and clogging water catchment facilities, such as reservoirs, with tons of sediment and debris. Thus fire effects vary both spatially and temporally. Smoke episodes from wild and prescribed fire also can lead to increases in respiratory afflictions, especially among susceptible infants and elderly persons with pulmonary sensitivities. Other off-site fire effects include the temporary closure of airports or highways because of visibility reductions caused by smoke, sometimes leading to traffic accidents.

In this chapter we explore some of the major global problems, as well as issues unique to North America. Solutions to problems are presented where possible, or if known. In other instances, the various sides to unresolved issues are discussed so that you may be able to draw your own conclusions. Forest fire issues are often complex and variable, so acceptable solutions can sometimes be elusive. As with most complex issues, generalizations must be avoided under most circumstances.

A forest fire typically sets in motion a complex set of reactions, depending on location, antecedent weather, and values at risk. Lightning ignitions in remote wilderness areas far away from human population centers may attract little attention; by contrast, a wind-driven fire that threatens homes or communities in the midst of a prolonged drought will require an immediate response. In the latter case, initial attack crews from a variety of government, volunteer, and even private and penal firefighters may converge on the incident, depending on jurisdiction and dispatch priorities (see Chapter 5, regarding fire suppression). If the fire exceeds initial capabilities, local officials may request assistance in the form of firefighting resources from regional or national coordination centers, which in extreme cases may involve branches of the military (for example, the Yellowstone fires of 1988).

In addition to firefighter responses, a bevy of community resources may be mobilized. For example, the Red Cross may set up emergency evacuation centers for homeowners displaced by a threatening fire. The Federal Emergency Management Agency (FEMA) and insurance adjusters will become involved, especially if the incident receives "disaster" designation. Hospital emergency rooms may be pressed into action to respond to a wide variety of symptoms: skin burns, ankle injuries, poison oak rashes, inflammations suffered by firefighters, or reduced pulmonary function by asthmatic residents who cannot tolerate exposure to smoke, among others. Local animal shelters may be called upon to board animals whose pens or stables are threatened by the approaching flames. Road closures and transportation delays will disrupt local commercial activities, including businesses relying on tourism revenues or airports requiring clear visibility corridors. Municipal water districts may plan and implement flood-control measures.

With all this activity, can we conclude that all forest fires are bad? Believe it or not, that question cannot be answered simply. Certainly, fires that cause excessive resource damage or any loss of life are highly regrettable. However, forests have burned for millennia, and many of the problems associated with wildland fire have arisen only since human settlement of formerly wild areas. Furthermore, a preponderance of scientific research into the ecological effects of wildland fires supports the notion that fires play an important and inevitable role in sustaining forests, shrublands, and grasslands. This body of evidence suggests that attempts to exclude fire from an area only postpone the inevitable. Eventually the area will burn, and perhaps more explosively than before—especially if fuels have built up from the suppression of low-intensity fires that previously would have removed understory vegetation and accumulations of surface fuel. Also, when appropriately harnessed, fires are a powerful and important tool for manipulating landscapes to serve human needs. Indeed, fire can be an extremely useful tool to the land manager. The intentional use of fire to achieve management objectives, otherwise known as prescribed fire, involves reduction of fuel hazards such as small trees and surface fuels accumulating on the forest floor. Other land management objectives (among many) for prescribed fire include improvement of wildlife habitat and creation of a favorable seedbed for plants requiring a bare mineral soil for seed regeneration (see Chapter 5). Thus many in our society view fire as possessing a split personality; it can be both a vexing enemy and a dutiful servant.

The disparity in views toward fire shows up as well with fire stakeholders. Public agencies with fire management responsibilities have different policies toward fire, and these may differ from the views expressed by private interest groups (see Chapter 6) or individuals. Although many public land agencies engage in firefighting activities, policies differ among agencies. For example, those agencies whose mission includes multiple use of natural resources (for example, the U.S. Forest Service, Bureau of Land Management, or some state jurisdictions) may have policies different from those of agencies charged with ecosystem preservation or wildlife habitat maintenance (for example, the National Park Service and U.S. Fish and Wildlife Service). These differences show up in the ways in which fires are managed in respective jurisdictions—for example, because of differing agency missions and fire policies. If a fire burns, say, in an area designated for multiple uses (for example, recreation, timber production, or wildlife viewing), the responsible agency may attempt to keep the fire size as small as possible, in order to minimize the effects on natural resources and nearby human communities.

By contrast, an agency charged with sustaining natural plant communities may actually set or encourage some fires to burn, so long as management objectives (that is, ecosystem restoration) are met without threatening human values. Fires that cross multiple jurisdictions can point out this conflict rather dramatically if agency missions differ. For example, the use of fire suppression equipment (such as bulldozers) might be restricted or even banned on a wildfire that spreads from a multiple-use area onto an adjacent national park concerned with preservation of ancient archeological resources. In those latter areas, the possible loss of irreplaceable artifacts or prehistoric treasures in the path of an oncoming fire would need to be weighed against the physical scars left by the heavy machinery used to fight fire. Often firefighters are constrained to fight fires with a "light hand on the land"—or using minimal-impact suppression techniques to reduce the possible disturbances caused by fire suppression operations.

To develop an appreciation for the problems, controversies, and possible solutions confronting forest fire managers and society in general, let's analyze in greater detail the issues that arise in the characteristic fires identified in Chapter 1 (wildfires that strike homes, killer fires, park and wilderness fires, friendly fires, and other classic fires). This classification scheme is my own contrivance and admittedly not standard; yet it suggests the variety of fire problems encountered worldwide and the complexity of potential solutions. Furthermore, this scheme should suggest that different types of fire will require unique management responses and that simplistic solutions are inappropriate. By discussing these problem fires individually, readers should gain insight into the full range of issues associated with managing wildfires and their impacts.

Wildfires that Strike Home

As noted in Chapter 1, the 1991 Tunnel fire represents a classic example of a type of problem fire that is bound to occur more frequently in the future as homes proliferate in the so-called wildland urban interface (also known as urban interface, urban intermix, or exurbia). The Tunnel fire (sometimes called the Oakland/Berkeley Hills fire) ranks as the worst single wildland fire in California history. Although the 2003 southern California wildfires burned more total area, no single wildfire has caused more damage than the Tunnel fire.

Fire is no stranger to the East Bay hills of northern California, where the native shrubs and eucalyptus trees (imported from Australia during the early 1900s) can be highly flammable, especially when dry in late summer. In fact, in the fall of 1970, a fire deliberately set by a local park employee burned thirty-nine homes close to the area eventually torched by the Tunnel fire twenty-one years later. In 1923 a fire pushed by 60 mph east winds spread downhill and decimated fifty city blocks and 624 homes north of the University of California campus (Carle 2002). To this day, visitors to the north campus area note that most of the houses have been constructed with terra cotta (that is, nonflammable) roofs, an indirect legacy of the 1923 fire.

Fires are even more legendary in southern California, where the predominant shrubby fuels (chaparral) explode into flames with rhythmic regularity. The native chaparral vegetation includes many plant species that resprout from burned root crowns following fire; other plants regenerate from subterranean seeds that germinate only in response to intense heat that penetrates the soil profile from above the ground. These regenerative mechanisms (resprouting and seed germination) are stimulated by the explosive flammability of the plants, many of which possess oils, waxes, and resins (also known as ether extractives). In fact, Mutch (1970) speculates that high flammability is a unique fire adaptation that, along with sprouting and seeding mechanisms, allows the dominant vegetation to persist in fire-prone ecosystems, such as southern California chaparral or some pine forests. Species without these adaptations have little chance of surviving the periodic severe burning. Similar speculation has been posited regarding eucalypts in Australia and pines in the southeastern United States (Whelan 1995).

Typically, autumn in southern California brings the ominous Santa Ana winds, as air flows from areas of high atmospheric pressure in the Great Basin to low pressure over the Pacific Ocean, west of Los Angeles. Pushed by the Santa Anas (also known as devil winds, or simply, Santanas), ten wildfires in southern California burned some 304,000 hectares (750,000 acres), an area roughly half the size of Rhode Island, during a two-week burning period in 2003, with twenty-two fatalities and 3,500 homes or structures destroyed or damaged. The ABC Monday Night Football program was moved from San Diego to Tempe, Arizona, out of concern for the apparent absurdity of playing a nationally televised game while the nearby hills burned.

Every western U.S. state has experienced the destructive mix of rampant urban development commingled with wildland fuels that explode in flames during a drought. Here the homes themselves can become the fuels that carry the wildfire, especially if they were constructed with combustible building materials (for example, roofs, siding, and decks). Even when the traditional summer fire season ends in the West, houses may yet burn in other U.S. locations, such as Florida, North Carolina, and Michigan. By contrast, the number of houses lost in wildland fires is much lower in southern Spain, Italy, and Greece, where building materials are generally nonflammable. So, part of the problem with burned homes in the United States may be related to our choice of home construction materials. Significant U.S. wildland fires involving loss of houses and structures are noted in Chapter 3.

The 2002 fire season destroyed hundreds of homes and outbuildings while producing the largest fires ever recorded in Arizona and Colorado, almost 500,000 acres (202,429 ha), and 137,000 acres (55,466 ha), respectively. Oregon recorded its largest fire in more than a century (450,000 acres, or 182,186 ha), eclipsing the burned area mark established by the Tillamook fire of 1933 (311,000 acres, or 125,911 ha). Ironically, federal employees with questionable motives were implicated with setting the record fires in Arizona and Colorado. Although employee-set fires are not necessarily rooted in malicious intentions, motives are certainly suspicious when hundreds of homes are destroyed and lives are threatened. Formerly wild areas have become so populated with private residences that housing losses are an inevitable consequence of large fires in all but the most remote areas.

When first recognized as a burgeoning problem in the mid-twentieth century, urban interface communities were viewed as a double dilemma for fire managers: wildland fires could burn into developed areas, and fire could spread from burning homes into adjacent wild areas. Over time, fires originating in the forest have destroyed far more homes and caused more damage than structural fires that escape onto adjoining wildlands. This situation only worsens as more and more people escape metropolitan congestion in favor of an idyllic life in the woods. Often these same people possess poor knowledge of and lackadaisical attitudes concerning the risks of living in a fire-prone environment. Residents may have few incentives for creating defensible spaces around their homes (such as the use of flame-resistant construction materials, the removal of fuel, and installation of water sources in the vicinity). Furthermore, mountain residents are recipients of fire protection provided by the government, a possibly gratuitous subsidy financed largely by all taxpayers. This subsidy extends to homeowners who qualify for low-interest loans or cash grants to cover uninsured (or underinsured) losses when an area receives disaster designation following a wildfire.

Ironically, urban interface or exurban dwellers in the United States tend to be more individualistic and less accepting of governmental rule-making. Homeowners in these areas tend to prefer to be left alone, to settle the wild as they please, and to contend with the forces of nature on their own. The myth of frontier self-reliance persists in these areas, only to be conveniently discarded when fires, floods, or other natural disturbances rear their ugly heads and set in motion governmental rescue efforts (such as FEMA subsidies).

Urban interface areas present unique challenges to fire managers. Wildland firefighters are not trained to put out structural fires, which generally require a completely different set of tactics and firefighting equipment than are usually needed when fires burn forest fuels. The strategies and tactics for contending with a free-burning fire within a wildland environment bear little resemblance to those needed to deal with a home engulfed in flames, so forest firefighter knowledge and technologies are not easily interchanged with their counterparts in the structural fire sector. A firefighter is not trained to deal with a deadly flashover in a home, any more than is a structural firefighter to contend with a raging crown fire in the woods.

During a wildfire, homeowners may be chagrined to observe a fire engine bypassing their residences in favor of a more defensible, safer location. Most homeowners do not distinguish between wildland and structural firefighters and expect some sort of protection if their house is threatened in a forest fire. What's more, prior to a fire, many homeowners may not recognize that their property is at risk, constructing their homes in high-risk areas that have burned repeatedly in the past, using combustible building materials on sites thick with flammable vegetation. Others may not recognize that the roads to their secluded hideaways are substandard or unsafe for fire engine travel or mass evacuation.

When wildland firefighters are diverted to protect structures during a wildfire, they are essentially not available to contain the fire's perimeter growth elsewhere, thus driving up the costs of fire protection. Ironically, concerns for firefighter safety can lead to controversial confrontations between fire agencies and private landowners. If firefighters pull back from a burning area because of safety concerns, homes or other resources may be burned or damaged, essentially sacrificed in order to keep firefighters out of harm's way. The number of homeowner evacuations and homeowner displacements may increase. On the other hand, homes or resources can be lost when firefighters employ aggressive tactics known as backfiring (or backburning), typically a last resort to rob fuels from an oncoming flaming front using intentionally ignited fires. Such an operation must be timed carefully so that the intentionally ignited backfire will be drawn into the convection column of the main fire, thereby blocking spread in the direction of the backfire. However, backfiring can have unintended consequences if applied incorrectly, increasing the burned area and adding to the area of active flaming—thus complicating management problems. In either case (that is, providing for firefighter safety or aggressive backfiring), homeowners may take issue with the decisions made to manage a wildfire, especially if private residences and personal belongings are lost.

After the fire is out, homeowners may find that their FEMA loans or property insurance policies are insufficient to cover the actual losses sustained. In addition to losing everything and having to start all over again, burned-out homeowners may suffer additional disappointments if their insurance does not cover their replacement costs, or if existing coverage has not kept pace with higher labor and construction costs—or if inflation protection is inadequate to provide full protection. Additional disappointments may be in store if building codes have changed, and insurance coverage is insufficient to rebuild to the new or higher standards.

Partial Solutions: Safeguards from fire for communities at risk include use of nonflammable building materials for individual homes and vegetation clearances between homes and surrounding flammable vegetation prior to the occurrence of a wildfire. In the Black Forest of eastern Colorado, residents have for years participated in slash/mulch programs, whereby thinned, small-diameter trees can be exchanged for chipped mulch at a central processing location. Community partnerships, such as the Pikes Peak Fire Prevention Authority, have formed or have incorporated with the intent of making residential developments fire-safe. The appropriate pricing of insurance premiums is often mentioned as one possible means for encouraging urban interface dwellers to build and maintain defensible spaces around homes, although insurance providers may have insufficient incentive to participate. Homeowners need to pay attention to exact replacement-cost limits in their insurance policies (different from fair market value or sales price), and possibly increase those limits if losses may not be fully covered. Owners of older homes may want to procure building code upgrade coverage, in case home rebuilding would require expensive compliance with recent code changes. Local jurisdictions should discourage or restrict home building in fire-prone areas (also known as fire-plains), using appropriate zoning authorization. A similar approach has had some success in dealing with development in U.S. flood plains. Ultimately, these measures place more responsibility upon home-owners and county planners to educate themselves about fire risks and their mitigation.

As an alternative to massive fire evacuations, which have become commonplace in the United States, Australian homeowners are encouraged to "prepare, stay, and survive" wildfire incursions. Homeowners might look to the Australian example for guidance on risk and hazard mitigation measures. Websites and extension publications provide guidance for homeowners to construct defensible space around their dwellings—for example, Firewise (www.firewise.org/communities) and Dennis (1992). Interested homeowners can consult with local building commissions or refer to the National Fire Protection Association guidelines (NFPA 299) for standards aimed at increasing the odds that individual houses will survive a wildfire. Government employee malfeasance in setting fires requires heightened accountability, sensitivity to personnel dereliction, and enforcement of existing laws.

Killer Fires: Understanding the Fatalities

Firefighting takes place in a high-risk environment involving exposure to a variety of dangers. From the moment firefighters answer a fire call, they are exposed to numerous potential dangers, whether en route to the incident, on the actual fireline, or while involved in support activities. Firefighters are killed or injured in numerous ways, including burnovers, aerial and surface vehicle accidents, and inattention to warning signals of health risks. Mangan (1999) lists six categories for the 133 firefighter fatalities occurring in ninety-four separate events during the period 1990–1998 in the United States:

- Heart attacks
- Burnovers
- Falling snags (that is, dead trees)
- Vehicle accidents, including single- or multiple-collisions, and individuals struck by a moving transport
- Aircraft accidents, including fixed- or rotary-wing craft, as well as aircraft-related accidents occurring on the ground
- Miscellaneous, including training, medical, suicide, or other causes.

A fatality on a wildland fire usually triggers a special investigation to ascertain the causes and possible lessons to be learned. Investigative teams are usually convened as quickly as possible after a tragedy fire in order to take advantage of evidence and eyewitness accounts to the greatest extent possible. Even so, every fire presents unique challenges to incident managers, and firefighter fatalities persist despite lessons learned from the past. Although we focused on the fourteen deaths in the South Canyon fire in Chapter 1, firefighter fatalities have occurred in numerous other fires (Chapter 3). For example, the South Canyon fire prompted comparisons with the 1949 Mann Gulch fire (eleven smokejumper fatalities) in terms of coincidental topographic features (steep, mountainous terrain adjacent to a major river) and the elite status and firefighting experience of those overrun by the flames (Maclean 1999; Maclean 1992). Discussions about the Mann Gulch fire evoke additional controversy because the crew leader survived by diving into a burned area that he had intentionally lit off to escape the oncoming flames, a strategy largely ignored or apparently discounted by his own crewmembers, most of whom eventually perished trying to outrun the fire.

As might be expected, fatalities occur also among those who are not as well trained or prepared for demanding, on-the-ground firefighting assignments. During the period 1990–1998, the largest group of fatalities occurred among volunteer firefighters (forty-one), including eighteen vehicle deaths, seventeen heart attacks, five burns, and one electrocution (Mangan 1999). In the 2001 Thirty-mile fire, three of the four fatalities were U.S. Forest Service seasonal employees in their first year of employment, with minimal experience and training.

In the 1990 Dude fire, another classic example, six state employees perished in the fire shelters (portable, heat-reflecting tents carried by all firefighters) they had deployed in a last-ditch attempt to escape a sudden flare-up resulting from the collapse of the massive convection column above the fire. Although all firefighters on the fireline carry fire shelters, their use can be controversial, because they do not withstand the highest temperatures or the most perilous situations encountered on a raging fire. Also, an individual deploying a shelter needs to make a quick decision regarding its necessity and then take the required time to find or create a cleared area, remove the packaging, then climb inside and lie prone to the ground. As a consequence, some individuals may use the shelter incorrectly while others may eschew their use altogether. Shelters were deployed by some but not all of those who perished at South Canyon (1994); personal protective items may not have been required at the time of the Mann Gulch fire (1949).

Entrapments occur when personnel encounter sudden increases in fire intensity in which planned escape routes or safety zones are cut off, unavailable, or compromised (Munson and Mangan 2000). Entrapments may or may not result in shelter deployment or injury, so some are considered near-miss situations. During the period 1976–1999, a total of 1,692 firefighters were entrapped by fire on 240 separate incidents, including the largest deployment of fire shelters on a single incident (107 during the 1988 Canyon Creek fire). Accordingly, an estimated 2,330 near-misses and 23,300 unsafe acts/conditions were encountered in the same time period (ibid.).

Firefighters are indoctrinated in early training to "fight fire aggressively but provide for safety first" as part of their ten so-called standard orders. In fact, firefighter safety has been the top priority since the 1994 South Canyon fire review. New recruits receive forty hours of classroom and field training, including exposure to fire tools, safety, and the contributors to wildfire spread. Upon completion of the basic course, recruits receive a "red card," signifying competency to work on a suppression crew. Certification for supervisory positions requires additional experience and training, as the firefighter progresses to squad boss, then crew boss, and on to higher classifications. Yet fatalities and injuries continue to occur with disturbing regularity, even among the elite and best trained. Fatalities occur in part because fires can blow up without warning, undergoing a transformation from a quiescent, low-intensity surface fire to a raging inferno that can trap an unwary or weary firefighter.

Wilson (1977) indicates that the line between fatal and nonfatal fires is thinly drawn and depends on numerous factors, including human behavior. The common denominators among tragedy fires include the following:

- Relatively small fires or deceptively quiet sectors of large fires;
- Relatively light fuels, such as grass, herbs, and small shrubs;
- Unexpected shifts in wind direction or wind speed;
- Uphill fire runs.

Surprisingly, most fatalities do not occur in heavily timbered or shrub-covered areas that erupt suddenly into violent fire. Fatalities can occur there, but grass fires are also killers, especially when fanned by sudden, unexpected winds or wind shifts. Thus quiescent fires can prove more deadly if firefighters are caught unaware or are lulled into complacency by fatigue or inattention.

Partial solutions: The best way to prevent fire fatalities is to make sure that firefighters receive training in recognizing hazardous situations, including fuel and weather contributors that lead to sudden increases in fire intensity. Firefighters need to implement such training while in the field, and further, recognize when fatigue and stress may cloud their judgment. At such times, they should be relieved of decision-making responsibilities. Improved protective equipment would be helpful, though experienced firefighters recognize the greater importance of improved decision-making, including stronger oversight by overhead teams regarding crew assignments on the fireline; no home or natural resource is worth the loss of human life. Strict adherence to the ten standard fire orders (see Chapter 5) and recognition of the eighteen warning situations (Chapter 5) should, in theory, reduce casualties, although fatalities and accidents happen in spite of the best training. Workers on the fireline need always to make safety their highest priority, including the placement of lookouts, improved communications, awareness of escape routes to safety zones, and working always from secured, anchored positions. Improved communications may imply questioning of orders from superiors, not so much as a challenge to authority but to make sure that all workers understand their assignments and their right to work in a safe environment. Firefighters also need to be physically and mentally fit to cope with stressful assignments involving long and arduous hours.

Park and Wilderness Fires

Fires in park and wilderness areas will not always threaten homes, but they still occupy a special place in the hearts and minds of most people. People value parks and wildernesses for a variety of reasons, including the spectacular scenery and relative isolation from the crowds, pollution, and crime found in the city. Parks and wilderness areas also provide visitors with inspiration and scientists with insights into ecosystem processes that operate across wild landscapes. Thus fires in parks and wildernesses provide natural laboratories for observing fires that both consume and perpetuate the vegetation found in those areas.

As noted in Chapter 1, fire has been part of most landscapes for thousands, if not millions, of years. Natural fire sources include lightning, volcanoes, and other spontaneous combustion sources (such as coal seams or haystacks). Human activity is the other major source of forest fires; it falls into numerous categories, including arson, debris-burning, campfires, railroads, equipment, smoking, and children. Thus many human-caused fires may not be set with malicious intent, but may escape control on account of an errant firebrand or a lapse in attention. Still other fires may result from a long tradition of woods-burning. For example, Native Americans used fire extensively for a variety of purposes (see Chapter 3). In some parts of the Southern United States the tradition of woods-burning has been practiced for generations and the customs and procedures for firing the forest have been handed down to the male family head and have been accepted socially. More recently, human causes of fire have ranged from accidental (for example, escaped clearing fires from industrial activities) to malicious (for example, intentional arson). In fact, instances of arson often increase with heightened fire danger in densely populated areas such as southern California, New South Wales (Australia), Capetown (South Africa), Vancouver (British Columbia), or southern Italy.

The frequency and extent of historical fires have varied depending on fuel, weather and climate, and topographic variations, as well as the scope and breadth of human interventions on the land—that is, management activities. In some areas, such as the Sierra Nevada Mountain Range of California, fire was so prevalent before Euro-American settlement that many common plants exhibit fire-adaptations, such as thick bark and fire-stimulated flowering, sprouting, seed release or germination (McKelvey et al. 1996). Additionally, fire affected the dynamics of nutrient cycling and biomass accumulation and decomposition, working in synchrony with climate to create vegetation mosaics at different spatial scales, from the plant community or forest stand to entire landscapes. Because fire affects so many important ecological processes, twentieth-century fire-exclusionary practices have had widespread impacts, most of which are not completely understood (ibid.).

In the Sierra Nevada, median fire return intervals (length of time between recurrent fires) vary with aspect, elevation, and vegetation types, but fifty years or fewer is typical, based on fire scar studies (ibid.). Some areas may have burned as often as once a decade. In the lower-elevation foothills, the midelevation mixed conifer zone, and throughout the eastern side slopes dominated by Jeffrey pine (*Pinus jeffreyi*), median fire return intervals were less than twenty years (ibid.).

A similar pattern emerges throughout most North American forests. Prior to Euro-American settlement, fire played an important, if not essential, role in determining forest structure and composition, as well as maintaining a shifting mosaic of vegetation across the landscape. Fire frequency was generally greater in areas of lower elevation (that is, foothills) that were exposed to natural and human ignitions and where fuels were more often susceptible to burning. In those areas, recurrent fires periodically created patches of trees of differing sizes and ages, depending on the severity and extent of fire growth. With settlement, fire frequencies actually may have increased as timber was extracted from accessible areas and cut-over lands, caught fire, or was ignited intentionally. By the 1930s rampant timber extraction and subsequent burning reinforced fears of a looming timber famine, thus justifying governmental attempts to keep all fires as small as possible, or better, to exclude fire from the forest (see, for example, the 10 A.M. policy of the U.S. Forest Service, Chapter 3). Although the policies were well-intended, the effects of fire exclusionary efforts (that is, biomass accumulation, stand density, and crown canopy closure), especially in ecosystems that burned more frequently in presettlement times, have led to reconsideration of the 10 A.M. policy in these areas, particularly ponderosa pine (*P. ponderosa*) ecosystems in the western United States. In those systems, contemporary wildfires burn with greater severity than observed in the past, largely because of fuel accumulations that accompanied fire exclusionary practices during the twentieth century (Arno and Brown 1991; Agee 1993; Covington and Moore 1994; Mutch 1970; Pollet and Omi 2002).

At higher elevations, fires were historically less frequent, but periodic burning still left indelible imprints on the landscape in terms of forest structure, ecological processes, and the landscape mosaic. Many of these higher-elevation areas experienced reductions in fire ignitions and burn area during the twentieth century as well, though, to date, the influence of fire exclusionary practices is more subtle than in lower-elevation areas that historically burned more frequently. Fuels may not have accumulated on the forest floor as in the lower elevations, although spacing between tree crowns may have been lower. High-elevation forests are generally cooler and wetter than those at lower elevations, resulting in shorter growing seasons, less surface fuel accumulation, and consequently less pronounced fire seasons. Even so, McKelvey et al. (1996) suggest that higher-elevation forests in the Sierra Nevada experienced less burning in the twentieth century, most likely because of more efficient fire suppression efforts as crews took advantage of the cooler, wetter conditions to limit the extent of burning. Eventually, even high-elevation forests will burn and regenerate, especially following a prolonged drought, as evidenced in the 1988 Yellowstone fires (Chapter 1).

Thus higher-elevation forests are fire-adapted landscapes, albeit with longer return intervals than found at lower elevations. Some of these forests may not burn for centuries, yet the vegetation may exhibit unique fire adaptations, such as cone serotiny in lodgepole pine (*Pinus contorta*) forests. Serotinous cones on a live tree remain closed until opened by heat, such as that provided by a forest fire. The opened cones release prolific amounts of seed that take root in bare mineral soil, thus initiating the next generation forest. These forests may not have been subject to the fuel accumulations associated with fire exclusion in lower-elevation forests, although tree density (number of stems per hectare) and canopy cover (percentage) may have increased.

Throughout the first half of the twentieth century, the National Park Service (NPS) and its sibling federal agencies attempted strict fire control in most parks and wilderness areas. However, as large fires continued to burn, the policy of fire exclusion came under increased scrutiny. The NPS took the lead among federal agencies in experimenting with fire reintroduction in both low- and high-elevation wildlands, beginning with prescribed fires in Everglades National Park during the 1950s. Numerous high-visibility national parks in North America followed suit, including Yosemite, Sequoia-Kings Canyon, Yellowstone, Olympic, Rocky Mountain, Glacier, and Grand Canyon. Managers turned to increased use of intentional prescribed fire to manage a variety of lower-elevation habitats, and relied on natural fire ignition (lightning) in more remote, higher-elevation areas. In fact, today 85 percent or more of Sequoia and Kings Canyon national parks is designated as natural fire zones (Carle 2002)—primarily higher-elevation areas where lightning ignitions are allowed to burn until extinguished by rain, snow, or lack of fuel—essentially a fire herding strategy with only occasional intervention on a flank or as needed to ensure against escape.

The impetus for the change in NPS fire policy was a document known as the Leopold Report, written to evaluate wildlife management in parks (Leopold et al. 1963). Drawing on writings of the forty-niners who flocked to the Sierra Nevada in search of gold and other riches in the mid-nineteenth century, the Leopold Report described the changes in vegetation that had occurred during a century of land exploitation and fire exclusion. According to the report, open and parklike areas had been transformed into "dog-hair thickets" as a result of overprotection from natural fires. The report went on to suggest that national parks be managed to re-create vignettes of primitive America by manipulating vegetation with controlled fire, possibly preceded by other fuel-reduction techniques (such as tree-cutting) in areas that were too overgrown because of fire suppression.

Research into the fire regimes of high-elevation parks and wilderness areas led to experiments during the latter twentieth century that allowed fires to burn if ignited naturally—that is, by lightning. Resource management plans were drawn up, specifying the prescribed conditions under which such ignitions would be allowed to run their natural course, so long as other values were not threatened. By most accounts, this program of so-called prescribed natural fires was quite successful in restoring fire to park and wilderness landscapes (Carle 2002), although a few costly escapes did occur. In general, the fires allowed to burn were confined to predetermined areas before being extinguished by season-ending precipitation. Occasionally, however, fires allowed to burn have become monsters—especially if active flaming zones become exposed to high winds, as often occurs with the passage of dry cold fronts through an area. That possibility became painfully clear during the Yellowstone fires of 1988, but other, lesser-known fires have had similar notoriety. For example, the 1978 Ouzel fire in Rocky Mountain National Park made a spectacular run toward the town of Allenspark, Colorado, before being extinguished by autumn snowfall. The escape of the Ouzel prescribed natural fire led to a shut-down in prescribed natural fire programs across the NPS, as happened following the Yellowstone fires in 1988.

It is worth noting here that prescribed natural fires were formerly called "let burn fires" but were renamed for purposes of public relations; essentially, the NPS became uneasy with the inference that such fires were not being managed. The subsequent terminology, "prescribed natural fires," was dropped following the 1994 federal policy review, largely for administrative nomenclature and financial tracking purposes. Today these fires are called wildland fires used for resource benefit—a confusing reference aimed at reinforcing the notion that some fires may be allowed to burn with prespecified objectives and under certain weather conditions, although wildfires confer no resource benefits. Other descriptors for this practice (essentially of herding fires) include "natural prescribed fire" and "natural wildland fires," although the "let-burn" moniker is pretty tenacious and still shows up occasionally in the press.

The rationale behind allowing fires to burn in the backcountry is that they pose relatively little threat and provide an opportunity to allow fire to assume a more natural role, akin to that in pre–European settlement times. Furthermore, this policy allows for managers to use partial suppression strategies, say, to control the unruly or threatening flank of a fire, without having to resort to all-out extinguishment. If managed correctly, remote fires usually are of little concern until and unless human developments become threatened. Unfortunately, once homes are threatened, these fires typically have grown too large and are difficult to extinguish by any means. Thus, one of the challenges is to restore fire to these ecosystems relying solely upon natural ignitions, with sufficient safety margin to protect homes and valued natural resources. Sometimes this safety margin evaporates as a result of unanticipated winds or if expected rains fail to materialize.

Park and wilderness fires thus pose special challenges to decision-makers attempting to manage fires across a complex landscape. A hypothetical landscape may consist of some or all of the following features:

- low- to mid-elevation forests that burned recurrently in the past with low-intensity surface fires;
- higher-elevation forests that burned with infrequent high-severity fires or low-intensity surface fires;
- alpine forests that rarely burn at all;
- visitor centers, housing, or high-use recreation areas;
- wild areas valued for recreation, wildlife, or commercial enterprise.

Each of the above features requires different perspectives and fire management strategies. For example, in the low- to mid-elevation forest, managers may attempt to restore the historical role of low-intensity surface fires that burned relatively frequently, removing surface fuels and small-diameter trees. Prescribed fires may be set intentionally in such areas to mimic the effects of historical fires. In these areas the managers may aim to restore or recreate forest structures that mimic those preceding European settlement. In higher-elevation forests, management objectives might reflect a desire to herd or contain fires to restricted areas using natural barriers such as rivers, lakes, and rock outcrops. In other instances fires may be permitted to burn based on historical analyses showing that one or more season-ending precipitation events will likely occur before the fire reaches a high-valued resource that must be protected from fire. Total fire exclusion may be the goal in alpine areas where fire damage could persist for centuries, or in developed areas where visitor use is concentrated.

The variety of fire management strategies employed in parks and wilderness areas has elicited much public concern and scrutiny over the years, especially after the 1988 Yellowstone fires. Less publicized activities still elicit outcries, whether such activities be prescribed burning of giant sequoia groves or fuelbreak construction in Rocky Mountain National Park. Park and wilderness managers currently attempt to restore fire as an ecosystem process along with an assortment of other ambitious but sometimes contradictory missions, such as preservation of spectacular views, scientific research, stewardship over natural resources, public recreation, and financial stability for park concessions. Thus fire presents special challenges to park and wilderness managers, just as do crowds, pollution, and crime—except that fires also may be encouraged in selected circumstances.

Partial solutions: Parks and wilderness areas represent unique laboratories for understanding the historical role of fire in natural ecosystems, improving knowledge about fire effects, and learning how to restore fire as an ecological process in landscapes. Proper planning is required to cost-effectively blend the appropriate mix of fire suppression activities in developed and high-valued areas, along with desires for ecological restoration using intentional prescribed fire or natural fires in the backcountry. As a benchmark, plans should aim to restore or re-create landscapes that will be sustainable—that is, able to withstand extreme fire events, as did ecosystems that predated European settlement. Currently, lightning fires may be allowed to burn in national parks and national forests with approved fire management plans, although policies require the suppression of all human-caused ignitions. Ultimately, all fires (including the human-caused) should be judged on a case-by-case basis—they could be suppressed if justified by the values at risk, but allowed to burn if they meet management objectives. This practice actually has a precedent in the so-called designated controlled burn (DESCON) program applied in the southern United States during the 1970s. Accordingly, managers would be given flexibility to employ a range of alternatives for each fire, from passive monitoring to all-out suppression, regardless of ignition source. While that may be appealing from a theoretical standpoint, implementation of such a program would require consummate skill by managers and willing acceptance by the public—a challenging combination under most circumstances.

Worldwide, isolated examples exist where land reserves and natural areas are being managed to retain or sustain the integrity of ecosystems where fire has had a historical and essential role. Kakadu National Park in Australia and Kruger National Park in South Africa are outstanding examples. These parks can be described as managed for biodiversity objectives, where the intention is to retain the role of fire-sensitive organisms and sustain the ecological integrity of ecosystems by maintaining a balance between too little and too much fire. Most often, these areas are set aside with the recognition that our knowledge of fire effects on plant and animal species is quite crude. Still, these areas are critical for increasing our knowledge base and also for testing hypotheses about the relationships between fire and affected organisms. Extra care is required where rare, endangered, or charismatic species are involved.

Friendly Fires

The 2000 Cerro Grande fire (Chapter 1) dramatically illustrated the problems that can occur when a prescribed fire escapes control and burns into an urban interface area. Unfortunately, most people are not aware of the many prescribed fires that are conducted safely and that meet the objectives as intended. Mistakes do occur—see, for example, BLM (1999) and Simard et al. (1971) for reviews of two prescribed fire escapes. These, however, are decidedly in the minority and should not overshadow the overall need for reducing fuels and restoring ecosystems across a landscape. In fact, only 1 percent of the 3,784 NPS prescribed fires set in the two decades prior to Cerro Grande resulted in escapes (*Los Angeles Times* 2000, as noted in Carle 2002, p. 243). Although escaped prescribed fires pose difficult issues for all involved, from homeowners to land managers, the fact remains that fires cannot be excluded from the forest indefinitely.

Notwithstanding the posture of public and private agencies that annually engage in firefighting operations, there are growing numbers who see the inevitability of fire recurrence and the futility of repeated campaigns aimed at fire elimination. Schiff (1962), Pyne (2001), Carle (2002), and other investigators have examined historical battles between those who persist in believing that fires can be excluded and those who would overturn this dogma by instituting widespread prescribed fire programs. Such arguments persisted throughout the twentieth century, starting with those who favored "light burning" (a handful of scientists and an assortment of timber and rancher interests) versus various public agency heads (the best known being Gifford Pinchot of the U.S. Forest Service), who generally felt that the public was not sophisticated enough to distinguish between unwanted and desirable fires. Early in the twentieth century, foresters in California debated the merits of so-called light burning in the understory of merchantable timber stands. Viewed as a perpetuation of Native American burning practices by some, the practice found adherents among industrial foresters from the Southern Pacific Company in the central Sierra Nevada and the Red River Lumber Company in Shasta County (Pyne 1982; Agee 1993). Arguments against light burning included the killing of young trees and assertions that fire control was required for efficient timber production.

Meanwhile, proponents for prescribed burning would periodically raise questions about the effectiveness of fire suppression efforts, whether in the southern United States (for example, Chapman [1912]; Stoddard [1931]) or the western United States (for example, Weaver [1943]; Biswell [1989]; et al.). Carle (2002) describes the evolution of sentiments, noting in particular the long tradition of woods-burning in the Southern states. These and other studies critiqued public agency fire policies that attempted to exclude fire from areas previously characterized by low-intensity surface fires. Toward the latter half of the twentieth century, evidence was mounting that high-severity, sometimes horrific, crown fires were replacing the low-intensity fires in these systems.

Each successive "bad fire year" (such as the 1994, 2000, 2002, and 2003 fire seasons) reinforces the notion that twentieth-century fire exclusionary policies were misguided at best, and at worst a monumental governmental mistake—at least in some ecosystems. In each of those years firefighters watched helplessly as thousands of homes were destroyed throughout the western United States. Scientists and land managers generally agree that attempts to exclude fire have led to unsustainable fuel levels in some dry, temperate forest ecosystems. As a consequence, these forests today are choked with younger, shade-tolerant trees that provide a ladder for surface fires to climb into the tree crowns and become destructive crown fires. The federal government has responded with ambitious plans for thinning the nation's forests to get rid of small diameter trees, which are the most flammable.

The two most viable approaches to thinning the forest are prescribed fire and mechanical thinning with chainsaws or other equipment. Other methods—including chemical application of herbicides, biological agents (such as livestock grazers), and reliance on natural decomposition—are either too costly or politically inexpedient to be considered here. Prescribed fire, or the intentional application of fire to achieve land management objectives, is generally considered the lower cost and more natural alternative for thinning a forest. A well-executed prescribed fire is confined to a predetermined area and burns under planned environmental conditions (such as wind and fuel moisture, among the several elements specified in the fire prescription); in addition, it produces the desired effects. Typical prescribed burn objectives include reducing fuel hazards, improving wildlife habitat, and creating favorable seedbed conditions—for example, for plant species that require bare mineral soil for establishing seedlings.

As noted above, prescribed fire has a long and controversial history in the United States. One of the biggest problems with prescribed fire is the fear and risk of escape. Furthermore, smoke—even from prescribed fires that are successfully contained to predetermined boundaries—can create problems, not the least of which are reduced visibility and public health effects. Reduced visibility in nearby travel corridors (land and air) can restrict commerce and lead to vehicle collisions, as well as limiting recreational opportunities. Smoke also can harm individuals with respiratory problems such as asthma or emphysema. In fact, the smoke from prescribed fires may provide the greatest deterrent to widespread fire use, since airborne particulate concentrations are of concern to state health departments and federal regulators such as the Environmental Protection Agency (EPA). Research has shown that smoke production is greatest during the smoldering phases of combustion, after active flaming but prior to extinction. This smoldering phase generates the greatest particulate loads, so managers strive to minimize its extent by comprehensive mop-up or by burning only when moistures are highest.

Mechanical thinning of the forest provides more exact control over the trees killed and removed, without the risk of fire or smoke, but it is generally more costly than prescribed fire. Thinning operations may be precommercial or commercial, depending on the stand age and whether or not the cut trees have marketable value. Thinning methods and standards may vary, depending on objectives. Thinning from below (or low thinning) usually refers to removal of small-diameter trees or saplings, either by hand-crews with chain saws or mechanized equipment such as feller-bunchers. Overstory thinning (or thinning from above) may remove larger trees that form part of the forest canopy. In either case, the stems, branches, and limbs left in the forest are known as thinning slash, or slash, which can create a fuel hazard unless removed from the forest or piled and burned.

Forest fuels can be reduced either with thinning or prescribed fire, although both treatments generate considerable controversy. These treatments are not exact substitutes for each other, even though both can reduce fuel loads. Thinning requires one or more entries into a forest stand to cut out and dispose of the unwanted trees. Fuels lying on the forest floor may not be removed during thinning, even though they may constitute a major fire hazard, especially as a source for heat transfer from surface flames into tree crowns. Prescribed fire can remove the litter and small dead fuels from the forest floor surface and kill small trees, but its effects will be patchy and perhaps less controllable than those of thinning, whereby standards can be imposed for the removal or retention of individual trees.

In addition, most trees thinned from the forest have little commercial value—and milling and processing capacity for small-diameter materials may not be available in some areas (such as the Colorado Front Range, and other arid areas). Thus the removal only of small-diameter materials (the primary carriers of fire, and as a result the bulk of the fuel hazard in an area) creates another problem—that is, disposal of the useless biomass removed. Landfills are already at or near capacity in some areas, and handling/transport costs can be prohibitive. Electric power cogeneration with the biomass has been attempted in places such as northeastern California and Washington, but it usually requires a government subsidy or assurances of uninterrupted supply from public lands. Processing of forest biomass into livestock feed or pellets that can be burned in wood stoves has been suggested, but those are largely unproven at the large scales necessary to reduce all fuel hazards on public and private lands.

The disposal problem is compounded if thinning operators and contractors are unable to make enough profit to warrant bidding on hazard reduction projects on public lands or stewardship contracts on private lands. Thus incentives (such as permission to cut larger trees) are required to encourage a private contractor to bid on a thinning project, in order to generate the revenues necessary to offset costs and improve profit margins. This is a key provision in the Healthy Forests Restoration Act of 2003 (described below), although it has been controversial inasmuch as it is viewed as a veil for increased logging and road-building on public lands. On the other hand, there is little evidence to suggest that removal only of large trees will reduce fuel hazards or wildfire severity. To the contrary, wildfire hazard will not be abated if small trees and logging slash are left on site after removal of only the large trees.

Many question whether sufficient area can be treated by prescribed fire or thinning to reduce the area burned by future wildfires. Thus recent government plans for reducing fuels in our nation's forests have generated a multitude of detractors. Vocal critics maintain that the U.S. government moved too slowly in recognizing and responding to the fuels crisis in the nation's forests, citing the increasing incidence of large, high-severity fires during the latter part of the twentieth century. In response, some officials maintain that the pace of fuels remediation is hindered by pressure from environmental groups who protest thinning projects or mistrust fuel treatments as a guise for cutting timber. That type of thinking led to passage of the 2003 Healthy Forests Restoration Act. As signed by President George W. Bush, the Healthy Forests Restoration Act may restrict environmentalists seeking legal recourse to stall or block agency management plans that include thinning projects to reduce fire hazards (or even the salvage of burned-over forests). Courts have been instructed to limit judicial review of thinning projects where the possibility exists for unthinned forest to erupt into flames while these management plans are being contested. As mentioned above, the act includes incentives for private operators to cut larger trees and derive forest products from harvested trees on public lands to improve their profit margins.

Prior to the passage of the Healthy Forests Restoration Act, disputes had led to advocates for arbitrary tree diameter limits on trees that could be removed during thinning. For example, in the Southwest, an arbitrary 12-inch (30.5-cm) diameter upper limit had been proposed for thinning projects, ostensibly to keep larger trees from being removed. In the Sierra Nevada, an upper limit of 30 inches (76.2 cm) in diameter had been proposed as the upper limit for removal in proximity to communities, with a limit of 20 inches (50.8 cm) elsewhere. These contrived limits were largely artificial, since tree diameter may not reflect age or flammability, or be an ecological indicator of forest health. Rather, the limits usually represented compromises negotiated by public agencies with antilogging environmentalists who are concerned about commercial logging in public forests, or opposed to road building in pristine areas, or against the liquidation of old-growth forests.

It remains to be seen whether the 2003 Healthy Forests Restoration Act will encourage or discourage lawsuits regarding diameter limits and thinning projects. However, administration guidelines issued in Fall 2004 suggest a looming battleground over management of the nation's forests, with logging and wildfire hazard reduction among several issues of contention. In the interest of streamlining bureaucracy, the guidelines vest control over management decisions to local forest supervisors, suspend requirements for environmental review, and limit public comment. Contentious bickering among government agencies and environmental interest groups over the wildfire hazard reduction and forest restoration seems inevitable.

Debates between government agencies and environmental groups also will arise over postfire rehabilitation of burned areas, including salvage logging and removal of coarse woody debris (CWD) following a damaging wildfire. At issue here is the pace and method of restoring the forest following a wildfire, but the debate goes even deeper into the core philosophies governing the management of the national forests—including the suitability of earning a profit from the sale and use of public natural resources. Some environmentalists maintain that the profit motive should not apply to public forests; they would just as soon allow natural processes—that is, plant succession—to govern postfire vegetative recovery. Government foresters maintain that salvage logging is required to pay for the restoration costs or to provide incentive for private companies to bid on the salvaged lumber (since agencies such as the U.S. Forest Service do not maintain a workforce of loggers and must rely on the private sector to remove timber from public forest lands). In addition, governmental managers favor a more active restoration of the prefire forest, including tree removal and planting to regenerate the burned-over area. Environmentalists mistrust government capabilities to manage or restore old growth forests, alleging lack of experience, decades of mismanagement, and collusion with private industry. These differences spill over into debates about the desirability of quickly removing dead and dying trees before they become susceptible to insect attacks and rot, thereby limiting their commercial value. Environmentalists would rather see forest succession dictate conditions that will restore the forest, but further, they distrust managers who may allow the cutting of live as well as dead trees.

In truth, very little is known about the ecological role of coarse woody debris, including fire-killed trees, and their disposition following disturbances such as fire. Superficially, we know that more dead wood in the forest can contribute to increased flammability, but recent evidence suggests also that CWD may be important for the maintenance of nutrient and moisture pools, as well as providing refugia for organisms important to ecosystem processes. Controversies arise because large quantities of CWD may be salvaged from public forests for economic returns following a wildfire. Salvage advocates note that wood quality will deteriorate on account of decay, and that wood will lose commercial value if removal is delayed inordinately (say, beyond one year). Furthermore, standing dead trees may be breeding grounds for future insect outbreaks or contribute to future wildfire hazard. So salvage decisions involve numerous tradeoffs and disputes.

Other postfire rehabilitation concerns relate to the efficacy of burned area emergency rehabilitation (BAER) treatments. After a fire is extinguished, a team of specialists will map and assess the severity of burned areas in order to prioritize areas in need of recovery and soil stabilization efforts. Typical measures include aerial seeding, tree planting, tree felling (or straw wattle placement) across contours, mulching, and fireline rehabilitation. These activities are sometimes controversial, for several reasons. First, seed mixes may contain exotic plant species that can outcompete native vegetation for moisture and nutrients and eventually take over the site. Second, weeds and exotic seeds can be introduced to recent burns merely by increased vehicle traffic in the area. Lastly, there is insufficient evidence to suggest that some treatments, such as contour felling or mulching, actually stabilize soil and overland water flows.

Despite these debates over thinning of the prefire forest, salvage of the postfire stems, and BAER activities, there is little argument that the decades of fire exclusion have led to conditions where many Western forests are now too overgrown and choked with fuel. This unsustainable condition is especially prevalent in areas previously characterized by frequent, low-severity fires. The current fuel levels in these forests are too dangerous to allow widespread use of prescribed fire to thin the forest without prior thinning and removal of understory fuels. Without some type of preparatory treatment, the risk is too great that a fire will climb into tree crowns and escape control lines. Thus widespread application of prescribed fire will require that many areas receive one or two prior entries to make the forest safe for fire use. The result will be forests that are safer for all fires, both wild and prescribed. Otherwise, the overgrown forest is not sustainable, but simply a catastrophe waiting to happen.

In the meantime, the areas in need of treatment far exceed the capability of governments, private contractors, and volunteers to reduce fuel loads. With so much of a backlog for areas in need of treatment, a concerted effort is needed, involving cooperation between government and private entities, to reduce fuel hazards and restore fire to ecosystems from which it has been denied in the past century. Thus, rather than stifle prescribed fires, the Cerro Grande fire (Chapter 1) and the ensuing massive 2000 and 2002 fire seasons actually reinforced the need to treat fuel loads. The National Fire Plan (PL–06–291) was signed into law by President Clinton, initiating a long-term plan for improving fire management capabilities to reduce fuel hazards so as to protect communities and natural resources while safeguarding firefighters and the public.

The continuing Western drought and 2002/2003 fire seasons helped to promote President Bush's Healthy Forests Restoration Act, which essentially supported the objectives of the National Fire Plan while calling for accelerated thinning of the forest, including measures to relieve the deadlock between environmental interest groups and government foresters. Concerns have been raised that the Bush initiatives will bypass important environmental safeguards and favor private industry interests. So it remains to be seen if increases in fuel treatments and prescribed fire will pay off with reductions in the frequency of catastrophic wildfires.

Partial solutions: One of the biggest challenges facing land managers is to achieve an appropriate balance between fire control and fire use across all ecosystems. Fire suppression will be needed in high value areas, such as the urban interface. Proper planning and execution of prescribed fires as part of an integrated program for forest management is essential. Ultimately, the solution involves creation of forests where fire can be prescribed safely as a management tool, including strategic placement of fuelbreaks across the landscape. Fuelbreaks are areas where flammable vegetation is converted or restructured (that is, by thinning of forest crowns) to provide safe access for firefighters and to provide anchor points for firing operations on wild and prescribed fires. The effort to treat fuels and restore fire to ecosystems will need to be sustained over the long haul. That will require financial commitment to cover multiple treatments and ongoing cooperation by federal, state, and local governments with private interests, including reliance on the best available scientific knowledge to arrive at management decisions. Managers also need to recognize that fuel hazard reduction is not required in all ecosystems, especially those where extreme fire behavior is related more to weather and climate than fuel structure. Patience and willingness to tolerate occasional mistakes will be needed to undo the harm caused by past fire exclusionary practices. Collaborative learning and partnerships between public and private stakeholders may prove useful for reaching consensus about approaches for dealing with wildfire problems.

Other Significant Fire Problems

The four fire types mentioned above are not the only problem fires, although they capture a broad spectrum of concerns and issues. Still other types of problem fires deserve mention, for they strike at the heart of civilized societies around the globe—that is, at their storehouse of natural and historical resources. In the United States, as settlers moved westward during the nineteenth century, they continually encountered and created burned landscapes. Settlers learned from observing Native American firing practices that fire was a powerful tool for clearing lands; later, fire followed the ax as lands were cleared for their timber stores and burned to dispose of logging slash. As settlement progressed, wildfires became feared as threats to developed communities, but also because of the possible effects on natural resources, such as water and timber. From the Peshtigo fire of 1871, through the fires that ravaged the Pacific Northwest in 1902, to the Idaho-Montana conflagrations of 1910, the United States grew increasingly concerned about the threats to communities and a possible timber famine.

Originally, U.S. national forests were reserved for their watershed, and later, their timberland values. Thus for most of the twentieth century, timber/watershed fires were of primary concern to foresters, while most of the public was located too far distant to care very much about forest fires. The 1910 fires in Idaho and Montana occurred shortly after the national forest system was created (1905), and the infant forest service needed to prove its worthiness of the public trust to provide forest protection (Pyne 2001). Even though the U.S. Forest Service did not extinguish the 1910 fires single handedly, the magnitude of the fires and their aftermath essentially settled a controversy over whether USFS could manage fires in the forest reserves and laid the groundwork for attempts to exclude fire from the forest. Cooperative fire protection with states became a reality in 1924 (Clarke-McNary Act), and systematic fire protection was institutionalized in 1935 with adoption of the so-called 10 A.M. policy (see Chapter 4). For further details on the evolution of fire policies in the United States, see Pyne (1982) and Schiff (1962).

Today, public forestry no longer focuses solely on cost-efficient timber removals. As an offshoot of the environmental concerns expressed since the first Earth Day in 1970, today's managers cannot focus only on fire suppression to protect timber resources. Now they must focus on long-term ecosystem health and restoration. As timber harvest has declined from U.S. national forests since 1960, so the emphasis has shifted away from resource extraction to ecosystem sustainability.

Currently, 161 million ha of public lands in the United States are at risk of wildland fires that could compromise human safety and ecosystem integrity (Williams 2003), including 73 million ha (30 million acres) of national forest lands (Bosworth 2003). These estimates are based in part on fuel conditions in ecosystems from which fire has been excluded during the twentieth century (GAO 1999). Most of these lands are in ponderosa pine forests of the interior West, from the Sierra Nevadas and eastside Cascades to the Colorado Plateau, the Rockies, and Black Hills. Prior to European settlement, most of these forests burned with frequent, low-intensity surface fires that regulated fuel loadings while allowing the survival of mature, fire-resistant trees. Fire exclusion, wetter climate, and deferred management during the past several decades have combined to leave many forests more dense with younger, smaller, shade-tolerant trees. Historically these forests were kept relatively open by periodic surface fires—however, today these forests are overcrowded with trees and susceptible to stress from competition, insect attacks, and catastrophic wildfires

(Bosworth 2003). These forests burn with high severity crown fires that kill all the trees, even some that might have survived the low-intensity fires of the past. Fire severity is exacerbated following drought. So the problem becomes the excluded fire, the one that perhaps was easily suppressed in the twentieth century but eventually occurs following prolonged drought, burning with greater severity than ever witnessed by Native Americans.

Restoration of ecosystems to accommodate pre-European levels of fire has been proposed as one possible solution, especially for warm, dry, temperate forests. Successful programs have been initiated in Northern Arizona and southwestern Colorado, with expansion proposed to New Mexico and the Colorado Front Range. In the long-term, these programs aim to restore the biodiversity and fuel levels that permitted ecosystems to survive historical fires, in order to prevent the wildfire catastrophes that have run rampant throughout the West in recent years. These programs involve collaborative partnerships between public agencies, private landowners, academics, and the interested public to develop creative plans for managing fuels, cutting trees, producing usable forest products, and reducing bureaucratic gridlock while satisfying interest groups. Furthermore, coordination among the numerous public and private entities at local, state, and national levels can be challenging, especially since participants may not agree on implementation details or because of differences in political agendas.

A different type of problem has developed in forests that are valued for their historical or archeological treasures. For example, Mesa Verde National Park and Bandelier National Monument have instituted fire restoration programs with a special twist because of the archeological and historical features under protection. These parks were established to preserve insights to the way of life of the Anazasi people, who occupied the Four Corners region of the southwestern United States and then vanished around the twelfth century A.D. Recent fires in these areas have uncovered additional evidence of ancient settlements. Over time some of the settlements became overgrown with vegetation and were largely forgotten until exposed by recent wildfires. Fires in these parks have resulted in damage to some of the archeological resources, including discoloration of artifacts and spalling of rock art. At the same time, fires provide insights into Anasazi culture by removing overgrown vegetation and exposing settlements and tools previously undiscovered. Fires burning in areas valued for their archeological resources pose challenges for firefighters trying to avoid damage to artifacts or former dwellings. These sites sometimes vividly display the scars that firefighting equipment such as bulldozers can leave on the landscape. Furthermore, some scientists question the suitability of fire restoration efforts in the pinon-juniper forests at Mesa Verde and Bandelier, which typically burn with high severity. More important, fires that affect tribal cultures (both past and present) symbolize a type of fire from which we have much to learn. American Indian use of fire to manage their environment can teach us much about sustaining the forests from which we derive useful products.

With the passage of time fires have become increasingly challenging, as might be expected with our changing societal values and demographics. In a post-9/11 age, forest fire managers are involved even in discussions about homeland security issues. That is not surprising, as forest fires were identified as a security issue during World War II, followed by the Cold War during the 1950s. What may seem surprising is the entire range of national priorities that can touch wildland fire managers. Forest fire incident management teams are now called upon to assist in national emergencies ranging from terrorist bombings to oil spills to *Columbia* shuttle salvage efforts. During a time when government budgets are shrinking, the public fire manager is truly being asked to do more with fewer resources.

Partial solutions: Land managers of necessity must form effective partnerships with public, private, and other interested entities to carry out land stewardship responsibilities. Partnerships are needed because of the sheer enormity of fire problems as well as the growing recognition that no single entity is capable of addressing all of the issues satisfactorily. Effective partnerships have been formed at the local level to develop community strategies for contending with fire threats and their aftermaths between public and private agencies—such as implemented thinning and prescribed fire practices around Flagstaff, Arizona, and Durango, Colorado. Internationally, the Nature Conservancy's Fire Learning Network provides a useful model for restoring fire to ecosystems. Partnerships seem to work best when all participants are involved in defining the problem and developing a collective vision for future solutions, when adequate time is allowed for building trust, and when rewards and incentives are included.

Summary: Problems and Issues

The problem fires noted above suggest a litany of issues, challenges, and partial solutions, many of which are not unique but in fact cross over between the various types of fire. Thus effective management of fuels can be seen as part of reducing threats to communities and homes, as well as improving fire safety. Similarly, fuels management is essential for restoring the biodiversity of ecosystems where fire has been excluded. Ultimately, prescribed fire will be needed to maintain ecosystems that have been restored. In summary, the issues involved with problems discussed above include:

Urban Interface Issues

- In the urban interface, fires of increasing size and severity destroy more homes, including houses built with flammable materials (for example, shake roofs, wood siding, and deck materials).
- Housing developments will continue to expand into natural fire alleys (or fire-plains).
- Homeowners face increasing responsibility for their own protection—that is, creation of defensible spaces with cleared vegetation and water resources.
- Property insurance premium rates alone do not seem to provide sufficient incentives for homeowners to become more active in ensuring their own fire safety.
- Increasing conflicts may arise between land managers who wish to use fire as a tool and homeowners who are fearful of fire.
- Wildland firefighters are not equipped or trained to put out structural fires. By the same token, structural firefighters are not equipped or trained for wildland fires.

Fatality Fires Issues

- Fatalities mount as firefighters are placed in increasingly hazardous fuel situations;
- Failure to learn from the past regarding firefighter death and injury;

- Inadequacy of personal protective gear and firefighter safety training;
- Firefighters are entitled to work in a safe environment.

Park and Wilderness Fire Management Issues

- Park and wilderness management requires a landscape management approach, unlike anything we've tried before.
- The advantages of letting natural forest fires burn sometimes conflict with modern societal values (such as clean air and the risk of escapes).
- The feasibility and practicality of fire restoration across an entire landscape has not been established.

Prescribed Fire and Thinning Issues

- Prescribed fire escapes and smoke create unacceptable risks to society, despite potential benefits in terms of hazard reduction and ecological restoration.
- Disposal of surface fuels and small-diameter thinnings is problematic because of lack of milling and processing capacity, and absence of markets for usable products.
- Public trust of land managers is low: Some groups mistrust government thinning plans as being a guise for timber harvest.
- Agencies and environmentalists disagree about thinning standards.

Other Issues

- Although most people know that fires are a problem, solutions are complex and elusive.
- Reversing the effects of nearly 100 years of attempted fire exclusion will require time, money, and human capital. In the meantime, some forests are choked with fuels and will continue to burn catastrophically. Other ecosystems have always burned with high severity and fuel hazard reduction may not be required.

- Realistic land management policies that include thinning and periodic burning to restore ecosystems will require commitments at the local, regional, and national levels.
- Allocation of fire management responsibilities among a myriad of federal, state, and local agencies along with private entities can be an enormous challenge.
- The role of fire in managing archeological resources is poorly defined.

These issues are inevitably complex and controversial, as they involve human values and complex natural ecosystems. A central question at the foundation of all the above issues is whether humans can truly manage free-burning wildfires effectively. Actually, that question is simplistic, since modern-day fire suppression forces are largely successful in suppressing most of the fires that occur in the forest. However, because of the inordinate damage caused by the remainder of a few destructive wildfires, the answer to this question has profound implications for agencies with fire suppression responsibilities. Typically these agencies spend billions of dollars per year attempting to put out fires. While few would question the need for fire suppression forces, the sheer magnitude and power of large, severe conflagrations can be overwhelming, seemingly defying any control efforts until the onset of favorable weather. Some would suggest that humans don't put fires out; rather, the argument continues, most fires are extinguished by weather changes that allow humans to "claim" the effectiveness of suppression activities. Actually, in any year, fire suppression forces are successful in extinguishing 97 to 98 percent of the ignitions that occur in wild areas, although the 2 or 3 percent that escape control may cause most of the damage. A related question relates to whether fires can actually be prevented. Here again, the issue is complicated because we never can know with certainty that a fire has been prevented from occurring. Furthermore, Smokey Bear promotions for years have proclaimed that only humans can prevent forest fires (which is largely incorrect, especially where lightning ignitions predominate). What we do know is that management activities can reduce the spread and intensity of fires that do break out, for example by thinning out the forest with mechanical or prescribed fire activities. (See Smokey Bear program, Chapters 6 and 7.)

The issue of whether or not to suppress natural ignitions depends entirely on location, available fuels, time, and values at risk. For example, suppression activities are automatically initiated against fires that occur near homes or that threaten high-value resources (such as timber, wildlife habitat, water, range, and so forth). On the other hand, fires that occur in a remote wilderness area may be allowed to burn if such a decision is consistent with an approved land management plan for the area. Usually such plans are based on ecological research that demonstrates the potential benefits to fire-adapted ecosystems in the area, buttressed by manager analyses that support the low likelihood of threats to high-value areas.

Solutions

The problem fires identified in this chapter share common roots and to a partial extent, similar solutions. For example, fires would not kill as many people or destroy nearly the number of homes they do if fuels were managed appropriately and the public were educated about the risks of living in fire-prone environments. Ironically, wildfires would not be nearly so devastating to forest and water resources if fires were allowed to assume a more natural role in ecosystems that were restored to some semblance of their historical or presettlement condition, in which periodic fires burned with low intensity and prevented excessive fuel buildup over widespread areas. Alternatively, some ecosystems (such as southern California chaparral) apparently burn more than ever and may require less fire, not more. Thus restoring the biodiversity of ecosystems, so that fire could resume a more natural role, would be a major catalyst for ensuring the sustainability of some, but not all forests.

Forest restoration alone is no panacea, and it is not needed everywhere. For example, the exclusion of fire from cold, wet, higher-elevation forests during the twentieth century has not created a problem with fuel accumulation. Those forests may owe their origin to fire, as the tree species are adapted to infrequent, high-severity fires that destroy the standing biomass, thus initiating a new forest. Furthermore, many forest health problems do not owe their origin to fire or its absence. Other culprits may be livestock grazing, overcutting, undercutting, and human population growth, just to name a few.

Fire seasons such as 1994, 2000, 2002, and 2003 in the western United States point out that the battle against forest fires cannot be won by continuing to pour money into fire suppression. What's more, our forests, people's homes, and firefighter safety are among the more obvious casualties of fire exclusion policies. Although well intended, the war against fire declared in the twentieth century is not winnable, at least in terms of its initial objective, of permanently ridding the forest of fire. As it is, some of our warm and dry temperate forests that previously were able to withstand periodic, low-severity fires have become clogged with fuel because of fire exclusion and have become susceptible to devastating wildfires following a random lightning strike or errant ignition. Firefighters are unable to put fires out in these clogged forests without assistance from the weather. Fire problems are most acute in the lower-elevation, dry forests that are also desirable living spaces for humans fleeing hectic cities. Fire protection and fuel hazard reduction are required in the urban interface, but away from populated areas, ecosystems may need restoration. At the same time, even in the urban interface the continued suppression of fires will eventually lead to a massive natural retaliation unless fuels and people are managed.

Ironically, part of the solution to all the problem fires discussed above is to restore the biodiversity and resilience of ecosystems so that they can once again tolerate fire, as they did before humans arrived on the scene. Yes, we must ensure that when forests burn (as they inevitably will), the fire will be of low enough intensity to cleanse the forest floor without roaring into the tree crowns and destroying everything in sight, as commonly occurs now. Such a restoration effort will be a huge and expensive undertaking, requiring massive amounts of patience and tolerance, and it may not rid the forests of devastating fires forever. But in the long run there is really no alternative if we want healthy, sustainable forests. A shift will be required in how we think about and manage our forests, as well as the homes that we live in and the air that we breathe. The public agencies that currently fight fires will need to learn how better to use fire, instead of trying to extinguish it. And those people and public interest groups that want to leave the woods alone will need to understand that some trees will need to be manipulated (that is, cut and removed, followed by a prescribed fire) in order to save and enjoy the forest.

The prevention of large, devastating wildfires will require a combination of active management approaches—the thinning of

forests and ecological restoration, fuel reductions and maintenance around homes and communities, and education programs
for homeowners and users of fire-prone areas. In other areas,
passive management activities, such as removal of active grazing
leases, may be appropriate in order to restore ecosystem
processes. Similarly, cooperation among myriad agencies,
groups of people, and tools will be required, some of which are
discussed in Chapters 6 (Agencies and Groups) and 7 (Print and
Nonprint Resources).

Eventually, the zoning of landscapes into discrete units typi-
fied by different fire management and land-use strategies may be
needed. Actually, some municipalities (such as Capetown, South
Africa, and Vancouver, British Columbia) use zoning to restrict
housing developments above certain elevations. Once zones have
been established for an area, different land-use priorities can be
implemented to develop management strategies for specific
zones. In some zones, aggressive initial attack might be employed
to protect houses, although development might be curtailed
there, or at least restricted, because of fire regime considerations,
the flammability of the vegetation, and historical spread patterns.
In other areas, forest restoration activities such as thinning and
prescribed burns might be employed.

A hypothetical landscape might be laid out as in Figure 2–1.
For illustration purposes only, suppose that an interstate freeway
passes through the eastern side of the fire management area
(FMA), with convenient access to Emerald Lake and the casinos
in Greentown. A two-lane highway bypasses the only other financial center, Bluetown, on the western side of the area. Several
tracts of commercial forest are located throughout the area. Un-
paved county roads (not shown) connect the two major roads, except through the Lonesome Peak wilderness area (roadless).

A well-conceived fire management plan might divide the
area into seven discrete units for planning purposes, based primarily on fire regime type (for example: short-interval surface
fires; long-return-interval, high-severity fires; mixed severity
types), fine-tuned with available fire history records, fuel inventory and computerized fire simulation runs, and local insurance
tables to rate the risks, hazards, and values in the seven units. Priorities would be established for increased fire preparedness (for
example, fire prevention, fuels management, pre-positioned dispatch) in the developed and high-valued management units
within the FMA. Full suppression with prioritized fuel treatments

Figure 2-1
Hypothetical Fire Management Area
A hypothetical fire management area is zoned into seven management units, including two urban interface units (1 and 7), four commercial units (2, 3, 5, and 6), and one wilderness area unit (4). Only major roads through and by financial centers Greentown and Bluetown are shown.

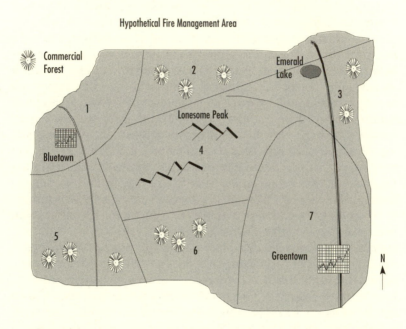

Hypothetical Fire Management Area

might be planned for the urban interface units (1 and 7) to protect homes and quality of life. Fuelbreaks might be planned in strategic areas to protect developments in the urban interface areas. Restoration of forest biodiversity, including thinning and prescribed fire, might be planned for some of the commercial forest units (2, 3, 5, and 6), especially where fire has been excluded from areas characterized historically by low-intensity surface fires. Recreation, hunting, and fishing might be promoted along with timber harvests in those zones. A monitoring strategy might be planned for the high-elevation wilderness area (unit 4).

Mere partitioning of the forest into fire restoration, fire use, and urban interface zones alone will be important but insufficient for assuring the best fire management practices. The remaining challenge will be to ensure the appropriate mix between desired fires (prescribed and perhaps natural ignitions) and undesired fires (which will be aggressively attacked) in different zones of the landscape. Special care will be required in managing fuels in boundary areas between units, to make sure that fires can be contained when desired. In management units where few people live (such as Unit 4), perhaps all fires might be potentially allowed to burn subject to prescribed guidelines, regardless of ignition source. Fuel levels in commercial forest zones (Units 2, 3, 5, and 6) would be managed so that prescribed fires could be safely implemented under all but the most extreme weather conditions, in which case no fires would be intentionally lit or permitted. In the urban interface zones (Units 1 and 7), homeowners would be educated about defensible space and encouraged to participate in community fuel management partnerships with fire management agencies. Fires would be suppressed aggressively near developments, but thinning and prescribed fire would be relied upon to manage fuels elsewhere in the urban interface zones. Although simplistic by design, the intent of the zoning exercise would be to make sure that the mix of prescribed and aggressively fought fires was appropriate to each locality.

In some parts of the world, all fires are suppressed without question—so all fires are considered harmful. However, in the United States we have evolved a mixed or nuanced view toward fire in the forest, one that calls for coexistence between fire suppression (for undesired fires) and fire use (for desired fires). Thus fire management options for a particular region or landscape can range from aggressive fire suppression to protect homes and high-valued natural resources, to the intentional use of fire (or allowance of natural ignitions) in parks and wilderness areas. Ultimately, in a strategically managed forest, every ignition will be evaluated with regard to its potential for harm versus benefit, regardless of ignition source. If deemed harmful, the ignition will be suppressed consistent with traditional practices; if considered useful, the ignition will be allowed to burn. A new set of principles and practices will be required. These include: (1) revised rating systems for fire danger and stand condition that govern the use of fire; (2) a national coordination system that mobilizes for fire use opportunities; (3) improved contingency planning that capitalizes on windows of opportunity in different geographic regions; and (4) prioritized treatment areas based on current condition and partnership opportunities (Williams 2004). We are a long way from having the sophistication and tools necessary to make such decisions, so in the meantime, we may need to content our-selves with making sure that we effectively manage the forest, fires, and people most of the time.

Ultimately, solutions to the wildfire problems identified in this chapter will most likely require a collaborative approach in-volving a mix of land managers, public interest groups, scientists, and academicians, among others. Although most agree on the need for science-based solutions, proposed activities also must meet the test of public acceptance. Thus consensuses will need to be built among all interested parties regarding: (1) the role of fire in maintaining and sustaining ecosystems; (2) the desired condi-tions of future fire-prone landscapes; (3) the mix of fire and other tools to be used in achieving those desired conditions; and (4) funding and willingness to pay for treatments. The complexity of issues and solutions will require government-private partnerships with healthy doses of accountability between participants. In the long run, the most successful partnerships will be those that rec-ognize the importance of adaptive management, or actions under-taken with the understanding and desire to learn by doing—in-cluding a willingness to tolerate and to learn from mistakes.

A final note relates to the role that mass media (television, newspapers, the Internet) can play in public perceptions about fire in the forests. During the summer, news media are filled with powerful images of forests aflame, many of which images rein-force typical negative stereotypes about fire and perpetuate a Smokey Bear mentality about forest destruction. These images may send an important message to forest residents and users, but at the same time, this message may oversimplify the effects and role of fire in our forests. Media messages in general focus only on the harm, with relatively little attention paid to possible improve-ments in ecosystem health and fire safety in well-managed forests. This bias is not surprising, inasmuch as the public may relate only to traditional messages about fire when they see spectacular flames, aerial retardant drops, and devastation to the forest and homes. In order to present more balance in media messages, new ways may be needed for thinking, talking, and reporting about fires. In short, we may need to develop a whole new vocabulary for communicating about the role of fires in our forests.

Literature Cited

Agee, J. K. 1993. *Fire Ecology of Pacific Northwest Forests*. Washington, DC: Island Press, 493 p.

Arno, S. F., and J. K. Brown. 1991. "Overcoming the Paradox in Managing Wildland Fire." *Western Wildlands* 17(1): 40–46.

Biswell, H. H. 1989. *Prescribed Burning in California Wildlands Vegetation Management*. Berkeley: University of California Press, 255 p.

BLM, 1999. Lowden Ranch Prescribed Fire Review. Final Report. USDI, Bureau of Land Management (also available at www.fire.blm.gov. /textdocuments/lowdenreview.pdf), 35 p.

Bosworth, D. 2003. "Fire and Forest Health: Our Future Is at Stake." *Fire Management Today* 63(2): 4–11.

Carle, D. 2002. *Burning Questions: America's Fight with Nature's Fire*. Westport, CT: Praeger, 298 p.

Chapman, H. H. 1912. "Forest fires and forestry in the Southern States." American Forests 18:510-517.

Covington, W. W., and M. M. Moore. 1994. "Southwestern Ponderosa Pine Forest Structure: Changes since Euro-American Settlement." *Journal of Forestry* 92(1): 39–47.

Dennis, F. C. 1992. "Creating Fire Safe Zones around Your Forested Homesite." Colorado State University, Cooperative Extension Service in Action, sheet no. 6.302.

GAO. 1999. "Western National Forests: A Cohesive Strategy Is Needed to Address Catastrophic Wildfire Threats." Washington, DC: US Government Accounting Office, Report to Subcommittee on Forests and Forest Health, Committee on Resources, House of Representatives. 60 p.

Leopold, A. S., S. A. Cain, C. M. Cottam, I. N. Gabrielson, and T. L. Kimball. 1963. "Study of Wildlife Problems in National Parks: Wildlife Management in the National Parks." Pp 28–38 in Transactions, 28th North American Wildlife and Natural Resources Conference. Washington, DC: Wildlife Management Institute.

Maclean, J. 1999. *Fire on the Mountain*. New York: William Morrow and Company.

Maclean, N. 1992. *Young Men and Fire*. Chicago: University of Chicago Press.

Mangan, R. 1999. "Wildland Firefighter Fatalities in the US, 1990–98." USDA Forest Service, Missoula Equipment Development Technology Center. Publication No. 9951–2811 MTDC, Missoula, MT. Available on MTDC Wildland Fire Safety Collection CD-ROM, Missoula Technology and Development Center, Missoula, MT.

McKelvey, K. S., C. N. Skinner, C. Chang, D. C. Erman, S. J. Husari, D. J. Parsons, J. W. van Wagtendonk, and C. P. Weatherspoon. 1996. "Sierra Nevada Ecosystem Project." Chapter 37 Pp. 1033–1040 in *Final Report to Congress*, Vol. II: *Assessments and Scientific Basis for Management Options*. Davis: University of California, Centers for Water and Wildland Resources.

Munson, S., and D. Mangan. 2000. "Wildland Firefighter Entrapments, 1976–1999." USDA Forest Service, Missoula Equipment Development Technology Center. Publication No. 0051–2853 MTDC, Missoula, MT. Available on MTDC Wildland Fire Safety Collection CD-ROM, Missoula Technology and Development Center, Missoula, MT.

Mutch, R. W. 1970. "Wildland Fires and Ecosystems: A Hypothesis." *Ecology* 51:1046–1051.

Parsons, J. W. van Wagtendonk, and C. P. Weatherspoon. 1996. "Sierra Nevada Ecosystem Project." Chapter 37 Pp. 1033–1040 in *Final Report to Congress,* Vol. II: *Assessments and Scientific Basis for Management Options.* Davis: University of California, Centers for Water and Wildland Resources.

Pollet, J. and P. N. Omi. 2002. Effect of thinning and prescribed burning on crown fire severity in ponderosa pine forests. International Journal of Wildland Fire 11(1):1-10.

————. 1994. "Fighting Fire with Prescribed Fire: A Return to Ecosystem Health." *Journal of Forestry* 92(11): 31–33.

Pyne, S. J. 1982. *"Fire in America: A Cultural History of Wildland and Rural Fire."* Princeton: Princeton University Press, 654 p.

————. 2001. *Year of the Fires: The Story of the Great Fires of 1910.* New York: Viking, 322 p.

Schiff, A. 1962. *Fire and Water: Scientific Heresy in the Forest Service.* Cambridge: Harvard University Press, 255 p.

Simard, A. J., D. A. Haines, R. W. Blank, and J. S. Frost. 1971. "The Mack Lake Fire." USDA Forest Service General Technical Report NC –83, 36 p.

Stoddard, H. L. 1931. *The Bobwhite Quail: Its Habits, Preservation and Increase.* New York: Charles Scribner's and Sons, 559 p.

Weaver, H. 1943. "Fire as an Ecological and Silvicultural Factor in the Ponderosa Pine Region of the Pacific Slope." *Journal of Forestry* 41: 7–15.

Whelan, R. J. 1995. *The Ecology of Fire.* Cambridge: Cambridge University Press, 346 p.

Williams, J. 2003. "Values, Tradeoffs, and Context: A Call for a Public Lands Policy Debate on the Management of Fire-dependent Ecosystems." In *Proc. of the 3rd International Wildland Fire Conference.* October 3–6, 2003. Sydney, Australia, 7 p. (CD-ROM).

————. 2004. "A Changing Fire Environment: The Task Ahead." Fire Management Today 64(4): 7–11.

Wilson, C. C. 1977. "Fatal and Near-fatal Forest Fires: The Common Denominators." *International Fire Chief* 43: 9–15.

3

Chronology

In this chapter we examine wildland fires and some of their documented effects. In the past, both natural and anthropogenic ignition sources contributed to forest fires. An increasing body of knowledge, bolstered in part by oral histories and anthropological inquiry, substantiates the use of fire by native peoples on all continents but Antarctica. However, we know that widespread fires burned on the planet even before anthropogenic ignitions. Charcoal, whether on old-growth trees, in the soil, or in cored-lake sediments, documents fire going back hundreds, if not thousands, of years. Fossilized charcoal dates fires back hundreds of millions of years, leading to speculation that fires may have been associated with the extinction of dinosaurs (Agee 1993).

We surmise that, more recently, native peoples fired the landscape repeatedly for a variety of reasons. The record of native firings is perhaps best documented in North America, Africa, and Australia, but evidence has been collected elsewhere as well (for example, Pyne 1997). In the United States, settlers encountered these fires and those of natural origin in the expanding frontier. Documentary records are more complete for more recent fires, so I will summarize the growing evidence for native firings in the forest, then trace documented large fires from pre-European times through the present in the United States.

Firing of the Forest by Native Americans

Martinson and Omi (2003) summarize findings from a meta-analysis of forty-seven North American fire history studies, documenting the widespread but clumped fire frequencies during the century preceding European settlement in North America. Each study was analyzed in terms of the average time span (in years) between successive fires (or mean fire-free periods) in forested ecosystems represented prior to settlement by Euro-Americans, using standard fire history techniques such as cross-dated tree rings, charcoal or pollen sediments, and stand reconstructions. The mean fire-free period ranged from 8 to 825 years, with a median of 33 years. No attempt was made in any of the studies to separate natural from human ignitions, but we are quite confident that Native American firings were substantial. Other literature sources (for example, Williams 1994, 2000; Pyne 1982) document the widespread use of fire by natives. The literature on Indian burning is extensive (see, for example, Barrett 1980; Lewis 1983; Gruell 1985) and growing. Furthermore, our estimates are conservative, relying solely on fire events that burned with sufficient intensity to be datable. Fires in grasslands and shrub fields are not included, although Native Americans may have ignited those vegetation types more than others.

Our knowledge that native firings were prevalent stems from a variety of sources, including textbooks and technical papers, websites, early newspapers, journals of early explorers, and oral accounts handed down through generations of American Indians. Naturally, the further back we go, the more we must rely on indirect evidence. Even though the study of anthropogenic fire has expanded in recent years, our ignorance always will be greater than our accumulated knowledge, especially relating to how flames first captivated the human imagination and then aroused utilitarian instincts (Omi 1990).

What's more, an abundant literature base documents the many reasons that Native Americans used fire in all regions of the United States after migrating from Asia 15,000 to 30,000 years ago (Phillips 1985). Of course, the incidence of fire in North America predates the Native American experience (Omi 1990). Lightning has ignited vegetation for millions of years (Stewart 1956). Thus

Native Americans discovered, but did not invent, wildland fires (Pyne 1982).

First Nations (in Canada) and native peoples elsewhere used fire for a variety of purposes. Documented reasons include the following (after Williams 1994):

- Hunting
- Crop management
- Improve growth/yield
- Fireproof areas
- Insect collection
- Pest management
- Warfare
- Economic extortion
- Clearing areas for travel
- Felling trees
- Clearing riparian areas

Still, the purposeful setting of fires by natives is difficult to document precisely (Williams 2000). Our knowledge of human firing practices prior to European settlement is thus based on informed speculation, relying on fire history studies and knowledge of how aborigines used fire in their daily lives. Also, this knowledge is evolutionary; we now generally believe that Indians burned far more areas than previously thought (ibid.).

European settlement dramatically changed the native firing patterns in North America (and by aboriginal peoples in Australia and elsewhere). Moreover, the eventual displacement of natives to reservations removed a powerful ignition source from wildland settings. As settlements were established, surrounding forests were seen as sources of necessary raw materials for construction and other uses, such as ship building. As forests were exploited and lands were cleared, massive fires in cutover areas became more commonplace. The practice of slash and burn—that is, clearing the forest and burning residuals (slash)—is practiced in developing cultures worldwide. When the forests are viewed as inexhaustible, these fires don't cause much alarm. But as forests are perceived to dwindle or go up in smoke, the greater becomes the need for creation of timberland reserves and centralized fire protection (and agencies such as the U.S. Forest Service). Efforts at fire protection soon follow, including overt attempts to exclude fire from the forest.

In the United States, continued settlement of formerly wild areas resulted in fragmentation of landscapes as roads, railroads, and waterways were developed. Forest vegetation was soon transformed, as the number and size of urban areas increased. Concurrently, higher populations increased the likelihood that fires would be set, while increasing the probability of earlier detection and subsequent suppression.

U.S. Wildland Fires and Their Impacts

Any listing of historical fires provides at best a crude snapshot of fire imprints over time. Dendrochronology, or the study and dating of tree rings, can extend our knowledge of fire in an area to several centuries before the present by relying on fire scarring of the cambium beneath a tree's bark. Techniques for dating fires rely on the dating of scars that form at the base of trees during a fire. Initially a triangular scar is formed when the heat from a fire persists around the tree bole long enough to penetrate the bark and kill the cambium, thereby creating a datable scar in the tree's growth rings (Agee 1993). Tree ring studies have been used to date fire scars in a variety of ecosystems in the southwestern United States (Swetnam and Betancourt 1990; et al.). Techniques have been developed for dating the year of fire occurrence and, to a lesser degree, for estimating the severity and extent of historic fires as well, thereby extending considerably the fire history for an area. By far, the overriding theme emerging from fire scar studies affirms the drop-off in fire frequency that accompanied Euro-American settlement of the North American continent, with the biggest reductions in lower-elevation forests. Similar drop-offs have been noted in Australia and elsewhere. Evidence for prehistoric fires has all but disappeared from European landscapes, and we can imagine that much of the global history of fire over the millennia remains undisclosed.

Although the historical record for fires becomes less certain as we look further back in time, we can be reasonably certain that the early white settlers in North America, Australia, and elsewhere encountered lands of prevalent burning. In particular, early European settlers in North America saw firsthand the impacts of natural fires, but they also learned to embrace native firing practices, which could be adapted to serve a range of settlement objectives, from pastoral pursuits of agrarian farmers to aggressive land clearing goals of the railroad, shipbuilding, and timber industries. In some cases, settlers may have simply expanded the firing practices from Europe (such as those invoked by shepherds from the Mediterranean region, or slash and burn cultures in Scandinavia); elsewhere, settlers adopted fire as a manipulative tool after observing the native use of fire.

Relatively little is known about the causes and extent of early fires, except as chronicled by nineteenth-century newspaper and travel accounts—or possibly, as can be inferred by the topographical location of fire scars on the landscape: lightning fires are most likely in mountainous terrain, particularly near ridges, just as human starts would be expected closer to human communities. These and other accounts, largely anecdotal but sometimes official, lead us to know that some of the earliest fires in the settlement of North America were large and devastating of life and human property, wiping out entire communities and probably underestimated in terms of their human casualties. For example, the Miramichi fire complex of 1825, actually started in New Brunswick, although its 3 million acres (1.2 million ha) included burned area in Maine also. The official death toll of 160 is widely considered an underestimate, omitting consideration of numerous others who attempted to evade the fire by jumping into the waters of the Miramichi River.

Tables 3-1 to 3-6 provide a historical perspective on fires that have directly or indirectly influenced public perception about fires occurring in U.S. wildland areas. Each fire is described generally by date (calendar year), name, location (state, with the exception of CA, which is divided into NCAL and SCAL), and significant impacts (size, structural damage, and fatalities). However, the impacts of each historical incident extend well beyond statistical description. By the time each fire was declared controlled, untold numbers of human lives and natural ecosystems may have been transformed, often dramatically if not tragically. Furthermore, many of the fires listed below led the way to changes in policy and new directions in fire and forest management (see Chapter 4). Thus the histories of forest fires in the United States, and elsewhere, provide insights into the evolution of societies in proximity to the incidents. At the same time, this chronology does not include countless other fires that burned prior to European settlement of wild areas in North America, that affected native cultures and ecosystems prior to recorded history, or that have simply escaped my scrutiny. Also, the significance of some of the fires represented here actually transcends the time period in which they occurred.

Frontier Settlement Fires

Fire was an essential tool as settler migrations displaced natives from their homelands, and as clearing the forest became important for agricultural and urban purposes. Inevitably, fires would break out as pioneers logged the forest for building materials and fuel. The Miramichi complex of fires (Table 3-1) followed a sequence typical for this time: logging and land-clearing leaving massive amounts of residual slash; and prolonged drought into autumn months, followed by intentional ignitions by settlers seizing opportunities to clear the slash (Pyne 1982). As demand for lumber outstripped supplies in the Northeast by the middle of the 19th century, a vast expansion of logging took place in the Lake States (Williams 1994), facilitated by improvements in railroad and water transport (for example, the Erie Canal). The geographic shift in logging focus was followed by the same pattern of igniting massive fires. The 1871 Peshtigo fire occurred at the same time as the more famous conflagration initiated by Mrs. O'Leary's cow in Chicago, killing over 1,200 people. Both fires were fueled by extreme weather conditions that stretched from the Ohio Valley to the Lake States to the Central Plains during October (Pyne 1982). High winds coalesced numerous logging and camping fires in northeast Wisconsin into one of the most damaging wildfires in U.S. history. The Hinckley fire (1894), like its predecessor the Peshtigo fire, was part of a 50-year era in which small logging towns were literally decimated by massive

Table 3-1
Representative Fires Occurring during the North American Frontier Settlement Period

Date	Fire Name	Location	Impact*	Fatalities
1825	Miramichi	ME, New Brunswick	3 million acres	160
1871	Peshtigo	WI/MI	1,000+ homes	~1,250
1894	Hinckley	MN	1,000+ homes	418
1902	Yacolt	WA	1 million acres	38

* Impact includes the size of the fire, the number of structures destroyed, and the untold financial loss.
Sources: Pyne 1982; Nobisso 2000.

fires. Nobisso (2000) recounts the heroics of an African American porter aboard a train trapped in the inferno one day during the Hinckley fire.

The Industrial Revolution led to a general feeling that human ingenuity, applied science, engineering, and mechanization could be used to overcome many societal problems, including forest fires. Further, by this time natives had been relocated to reservation lands, thus removing a potent ignition source for forest fires. The US Forest Service Organic Act was passed in 1897 to improve and protect forestlands, or to secure favorable conditions of water flows, and to furnish a continuous supply of timber from existing public reserves. These reserves originally were created starting in 1891 under the Department of Interior, with the hope of protecting watersheds and timberlands primarily from destruction by fire but also to promote other uses. In 1905 these reserves were transferred to the Department of Agriculture with its Bureau of Forestry, then renamed Forest Service. In 1907, the forest reserves were renamed as national forests (Steen 1992). Meanwhile, the ongoing fire seasons, particularly 1903 and 1908, burned large areas and thus posed continuing challenges for the infant forest service. Although timber management techniques had precedents in European forestry, there were few precedents for public agencies to manage fires over a large land area. Eventually, the Forest Service and public agencies would respond to this challenge by institutionalizing fire exclusion across the nation, but first it would need to confront those who favored a more natural and historic role for fire.

Early Years of Forest and Fire Management

Woods burning has a long tradition in the southern United States, where forests were customarily fired as part of local folk practices. Burning was thought to rid the forest of snakes, ticks, and boll weevils, while stimulating the green-up of grasses desired by grazers. It also provided a low-cost alternative for controlling the invasion of hardwoods and shrubs in timber stands. Whether superstition or observed fact, generations of Southerners are said to have burned the forest out of ritualistic tradition (Pyne 1982). So it is not surprising that early advocates for using fire in the forest hailed from the South, expounding the virtues of "light burning" as an alternative to the government's fire exclusionary approach to forest management. What's more, evidence began accumulating about certain plant species—such as longleaf pine (*Pinus palustris*)—and also animal species with adaptations for and even dependence on fire, thus adding ecological justification for light burning. However, the forces in favor of systematic fire control proved much more persuasive, aided by the specter of fires yet to burn in the backcountry.

In fact, the so-called light-burning controversy also had roots in California, where advocates (settlers and timber-owners) as early as 1880 saw the potential for periodic surface burning to reduce fuel hazards (ibid.). By contrast, government foresters saw light burning as a political threat to their agenda of reducing conflagrations through systematic fire control. Timber companies in northern California were also proponents of light surface burning, reasoning that fuel loads would otherwise build up in the forest and eventually lead to uncontrollable conflagrations.

During the period 1910–1929, the debate between light burning and systematic fire protection came to a head on several fronts, both scientific and cultural, even as major fires were burning up homes in the north-central and western United States (Table 3-2). Frederick Clements (1916) developed his "monoclimax" theory of plant succession, indicating that climate (and not disturbances such as fire) determined the developmental trajectory and endpoint of vegetation on a particular site over time (Agee 1993). By not including fire in his model of plant development, Clements indirectly contributed to support for the idea of European foresters (on which U.S. forestry was initially based)

Table 3-2
Representative Fires Occurring during the Early Years of Forest and Fire Management in the United States, 1910–1930

Date	Fire Name	Location	Impact*	Fatalities
1910	Great Idaho/Montana fires	ID/MT	3 million acres	85
1918	Cloquet	MN	~4,000 homes	551
1923	Berkeley	NCAL	584 homes	0
1929	Mill Valley	NCAL	117 homes	0

* Impact includes the size of the fire, the number of structures destroyed, and the untold financial loss.
Sources: Pyne 1982; Foote 1984.

that fire was somehow inimical to development of the climax forest, and therefore could be excluded.

Institutionalization of Fire Exclusion

In the United States, the policy of fire exclusion reached fruition in 1935 in the formulation of the 10 A.M. policy by the U.S. Forest Service, which justified the use of suppression forces as needed to extinguish all fires by 10 o'clock the following morning. The Matilija and Tillamook fires (Table 3-3) only added to the growing list of major fires providing rationale for instituting such a national policy. The fires that prompted this "experiment on a continental scale" (Pyne 1997, p. 195) included the numerous escapes from slash and burn activities, but the policy extended to all wildlands under the protection jurisdiction of the U.S. Forest Service. As a result of the policy, any fire that broke out essentially could be fought with the same moral imperative as a war with a foreign enemy.

The stock market crash of 1929 and vast pools of labor made available during the Great Depression provided additional rationale for putting people to work while embarking on a grand experiment to exclude fire from the forest. The massive construction efforts of the Civilian Conservation Corps included numerous fire control facilities across the country, including fire lookout towers, ranger stations, and firebreaks and fuelbreak systems.

Table 3-3
Representative Fires Occurring during the Time Period when Fire Exclusion was Institutionalized in the United States, 1931–1950

Date	Fire Name	Location	Impact*	Fatalities
1932	Matilija	SCAL	220,000 acres	0
1933	Tillamook	OR	311,000 acres	0
1936	Bandon	OR	386 buildings	13
1937	Blackwater	WY	1,100 acres	15
1941	Marshfield	MA	450 homes	0
1943	Southern California (incidents)	SCAL	200 homes	11
1947	Maine Forest Fire Disaster (incidents)	ME	1,200 buildings	16
1949	Mann Gulch	MT	5,000 acres	13

* Impact includes the size of the fire, the number of structures destroyed, and the untold financial loss.
Sources: Pyne 1982; Foote 1984; Wilson 1977.

The institutionalization of fire exclusionary practices in the United States did not occur without its detractors. In fact, the large, damaging wildfires of the early twentieth century had earlier spawned a spirited debate over the wisdom of attempting to keep fires from the forest (Carle 2002). The light-burning controversy, though officially discredited earlier, would continue to rear its head as major fire events could be interpreted as justification for both systematic fire protection and the need for fuel treatments such as prescribed fire (the modern-day equivalent to light burning).

Forest and Fire Policy Reflections

As the United States recovered from World War II and transitioned into its Cold War with the USSR, fire management became increasingly mechanized, relying on machinery and equipment to supplement labor-intensive fire operations for building firelines. Federal transfers of excess property provided states with access to vast stores of military surplus planes and trucks that could be converted to firefighting purposes. Across the United States concerns were expressed about real and perceived environmental degradation, resulting in landmark legislation aimed at protecting forests, creating wilderness, and maintaining roadless areas—all aimed at sustaining a variety of natural resource values. Meanwhile, the federal agencies with fire management responsibilities became increasingly concerned with environmental impacts associated with forest management decisions and operations (for additional details, see Chapter 4).

Increasing urbanization pushed city residents to seek solace and individual pieces of the American dream in forested and other formerly wild areas. Fires in the so-called urban-wildland (or exurban) interface became increasingly commonplace, with attendant damage to homes and commercial enterprises. The Bel-Air fire (Table 3-4) in particular provided a harbinger of fires yet to come, initiated in relatively remote wildland areas, then sweeping down into communities and gobbling up homes by the hundreds. Southern California developed a well-deserved reputation for typifying all that can go wrong when homes are built in the path of fires pushed by Santa Ana and coastal winds (from the east and west, respectively), while pulled by steep slopes covered with explosive vegetation. In such environments firefighters would sometimes lose the battle, as with eleven fatalities in the Loop fire in 1966.

Table 3-4
Representative Fires Occurring during the Time Period of Forest and Fire Policy Reflection, 1951–1970

Date	Fire Name	Location	Impact*	Fatalities
1953	Rattlesnake	NCAL		15
1956	Inaja	SCAL		11
1961	Bel-Air	SCAL	505 homes	0
1961	Harlow	NCAL	106 homes	2
1963	Staten Island	NY	100 homes	0
1964	Coyote	SCAL	106 homes	2
1964	Hanley, Nuns Canyon Fires	NCAL	295 homes	0
1966	Loop	SCAL		11
1967	Sundance	ID	50,000 acres	2

*Impact includes the size of the fire, the number of structures destroyed, and the untold financial loss.

Sources: Pyne 1982; Foote 1984; Wilson 1977; http://www.fire.ca.gov/FireEmergencyResponse/HistoricalStatistics/HistoricalStatistics.asp.

Environmentalism Takes Off

Earth Day 1970 ushered in an era of heightened environmental awareness, including passage of landmark legislation such as Clean Air, Clean Water, and Wilderness acts. In recognition of possible shortcomings of the 10 A.M. policy, the National Park Service experimented with its so-called Let Burn policy, whereby lightning ignitions in high-elevation remote areas would be allowed to burn until extinguished by natural causes—for example, winter rainstorms. Such ignitions would later be renamed as "prescribed natural fires" as a public relations strategy, but the Let Burn name still stuck and persists to this date. All public agencies started thinking about the increased use of prescribed fires to achieve land management objectives, although implementation lagged. In the southern United States, the U.S. Forest Service developed its designated controlled burn program (DESCON) whereby even human ignitions might be allowed to burn in certain areas if the benefits exceeded the cost of suppression.

This time period also saw several failures in implementing prescribed fire or fire use policies. The 1980 Mack Lake fire (Table 3-5) was intentionally ignited as a prescribed burn in order to remove logging debris and ultimately create habitat for the endangered Kirtland's warbler (*Dendroica kirtlandii*), which nests in Jack pine stands (Simard et al. 1983). Analogously, the 1978 Ouzel fire in Rocky Mountain National Park and some of the 1988 Yellowstone fires (Table 3-6) were allowed to burn but eventually became uncontrollable. Both those fires caused cessation of the National Park Service prescribed fire policies in place at the time. At various times public agencies have experimented with the idea of restoring fire in certain remote locations, allowing certain ignitions to burn if they meet predetermined environmental conditions (such as wind speed, fuel moisture, and drought indicators). The 1988 Yellowstone fires (Chapters 1 and 2) included several such ignitions, which started in adjacent national forests and were allowed to burn into the boundaries of the national park. These so-called prescribed natural fires, more recently called wildland fires for resource benefit, can be inherently risky, especially if they are permitted to burn and subsequently escape their maximum allowable burn area. The 1996 Oregon incidents illustrated some of these difficulties when five separate fires escaped control on national forest lands.

Although it is difficult to pinpoint the time with certainty, sometime around 1970 the arid western United States apparently entered the end phase of a prolonged wet climatic period characterized by unprecedented growth and density of forests during a period of deferred management caused in part by environmentalist appeals to agency management plans (Bosworth 2003). As

Table 3-5
Representative Fires and Their Effects during the Time Period of Increasing Environmentalism, 1971–1990

Date	Fire Name	Location	Impact*	Fatalities
1970	Wright	SCAL	103 homes	0
1970	Laguna	SCAL	175,000 acres	
			382 homes	5
1970	Wenatchee	WA	200,000 acres	1
1976	Battlement Creek	CO		4
1977	Kanan	SCAL	224 homes	1
1977	Sycamore	SCAL	234 homes	0
1980	Mack Lake †	MI	24,000 acres	
			44 homes	1
1980	Panorama	SCAL	325 homes	4
1985	49er	NCAL	148 homes	0
1988	Yellowstone †	WY/MT	2.4 million acres	1

*Impact includes the size of the fire, the number of structures destroyed, and the untold financial loss.

† Prescribed burn.

Sources: Pyne 1982; Foote 1984; Simard et al. 1983; Wilson 1977; http://www.fire.ca.gov/FireEmergencyResponse/HistoricalStatistics/HistoricalStatistics.asp.

the Western region transitioned to a relatively dry period prior to the turn of the twenty-first century, forests became ripe for the large, catastrophic fires of the past few years.

Increasing Safety Concerns

Wildfires became a greater national priority with the onset of the twenty-first century, even as firefighting became a colossal multinational industry. Suppressing fires now meant the mobilization of massive amounts of suppression resources to the site of a fire, including sophisticated remote sensing and telecommunications gear to assist crews and heavy equipment on the ground. Costs and losses continue to mount, and it seems almost preordained that fires should have gotten larger, killed more firefighters, and destroyed more homes in response to the management mistakes of past and present. In fact, fire management efforts have been overwhelmed by unprecedented levels of available fuel, with a big assist from episodes of prolonged drought in the southern and western United States, following on the heels of two decades of unprecedented moisture.

The four fatalities suffered in the 2001 Thirtymile fire (Table 3-6) were significant in that they occurred despite the substantial revisions in fire policy (emphasizing firefighter safety) brought about by the tragic 1994 South Canyon fire (Chapters 1 and 2). The Cerro Grande and Valley Complex (2000) were among the numerous incidents that led to the creation of the National Fire Plan. This plan called for the expenditure of $1.6 billion allocated to the following areas:

- Firefighting capability: Ensuring adequate preparedness for future fire seasons. Agencies funded at 100 percent of their most efficient level (see Chapter 5), including 8,000 new hires
- Rehabilitation of burned areas and landscape restoration
- Investment in projects to reduce fire risk (through fuel hazard reduction)
- Community assistance: Work directly with communities to ensure adequate protection in the urban interface
- Accountability, plus adequate oversight and monitoring

Table 3-6
Representative Fires Occurring during the Time Period of Increasing Concern for Firefighter and Public Safety, 1991–present

Date	Fire Name	Location	Impact*	Fatalities
1990	Paint	SCAL	479 homes	1
1990	Bedford	SCAL	120+ homes	0
1990	Dude Creek	AZ	28,00 acres Numerous homes	6
1991	Tunnel	NCAL	2,103 homes	25
1992	Fountain	NCAL	307 homes	0
1993	Old Topanga	SCAL	350 homes	3
1993	Laguna Beach	SCAL	366 homes	0
1993	Altadena	SCAL	118 homes	0
1994	South Canyon	CO	1,200 acres	14
1996	Oregon (incidents)	OR	5 escaped prescribed natural fires	0
2000	Cerro Grande†	NM	47,000 acres 235 houses, (400+ residences)	0
2000	Valley Complex	MT	100,000 acres 100+ structures	
2001	Thirtymile	WA		4
2002	Hayman	CO	137,000 acres 131 homes 466 outbuildings 1 commercial building	5
2002	Rodeo-Chediski	AZ	460,000 acres 423 homes	
2002	Biscuit	OR	499,570 acres $134 million	
2003	Southern California	SCAL	750,000 acres 3,500 homes, $2B	22

* Impact indicates the size of the fire, the number of structures destroyed, and the estimated financial loss.
† Prescribed burn.
Note: The five fatalities in the Hayman fire occurred in a vehicle accident while driving home from the incident.
Sources: www.nifc.gov; http://www.fire.ca.gov/FireEmergencyResponse/HistoricalStatistics/HistoricalStatistics.asp.

Several states suffered through record fire seasons during this time period. Florida endured a record fire season in 1998, with over 2,200 wildfires burning nearly half a million acres. At one time all residents of an entire county were evacuated.

The Rodeo-Chediski fire of 2002 burned nearly 500,000 acres (202,429 ha) in less than a month, the largest wildland fire in Arizona's history. Many factors contributed to this fire, most prominently an extremely dry summer within a five-year-long drought in the Southwest that continues to date. Human carelessness and greed also contributed, as the fire was intentionally set by a government employee apparently lured by the prospect of the higher wages to be garnered by participation in suppression activities. Other contributors include a buildup of fuel loads in the forest and contentious political issues such as thinning, logging, and road-building.

The 2002 Biscuit fire in the Siskiyou National Forest was the largest in Oregon history, burning more than 499,000 acres (202,024 ha), including limited spread into northern California. The Oregon fire season was the worst in more than 140 years, with costs of over $150 million for the Biscuit fire alone. Although fire is a natural part of the Pacific Northwest environment, there is broad interest in making it less destructive and less harmful to people, their property, and forested landscapes.

The 2002 Hayman fire was the largest in Colorado history, spreading over four counties (Park, Jefferson, Douglas, and Teller counties). Following a prolonged drought, the fire was set by a government employee apparently attempting to dispose of a letter from an estranged spouse—despite a fire ban because of the dry conditions. The fire burned 137,760 acres (55,773 ha) during June and July, destroying or damaging 600 structures (including 133 residences, 1 commercial building, and 466 outbuildings), with a total suppression cost of more than $39 million.

In 2003, the Cedar fire became the largest fire in California history, burning east of San Diego as part of a complex of fires in the southern part of the state during October. A lost hunter in the Cleveland National Forest apparently set the fire, after he fired a flare in an attempt to alert potential rescuers to his location. Air tankers in the vicinity were not dispatched, because of the looming darkness and concerns for pilot safety amid the high Santa Ana winds. That fire alone contributed to one firefighter and thirteen citizen deaths, more than 2,000 residences destroyed or damaged, and over 280,000 acres (113,360 ha) burned.

As of this writing, the 2003 southern California fires may be relatively fresh in the minds of most readers. But the record summarized in Tables 3–1 through 3–6 shows that southern California has experienced numerous historical fires involving massive destruction of lives, homes, and property. Fortunately, no recent fires rival the extent or fatalities of fires of the frontier settlement period (Table 3–1). However, today's wildland fires are much more damaging, given the concentration of high-valued resources such as trophy homes in the urban interface. No single wildland fire in modern times has been as destructive as the Tunnel fire in northern California (see Chapters 1 and 2) and the 2003 southern California conflagrations. Fortunately for the homeowners in those areas, most losses were insured or at least insurable, even though coverage often falls short of actual replacement cost. Compared with other California natural disasters, no single wildland fire complex or event exceeds the uninsured losses of the 1906 San Francisco earthquake and subsequent fires.

Although the years 2000 and 2002 produced some of the worst cumulative fire seasons in more than fifty years (in terms of total acres burned), individual fire events such as the 2003 southern California fires have punctuated the nation's fire history. In the short term there really is no plausible reason to expect that such incidents will decrease in their frequency or severity.

In the long term, however, there is reason for guarded optimism that measures such as the 2003 Healthy Forests Restoration Act (see Chapter 4) will provide relief from the litany of recent catastrophic fires, devastated communities, and damage to forests and watersheds. Also, funded research and collaborative endeavors under the National Fire Plan and Joint Fire Science Program (Chapter 6) should improve our understanding of fire and fuel dynamics as well as provide opportunities for creative partnerships between all affected parties. Ultimately, though, all of these programs will have been for naught unless the U.S. public faces up to the realities of living in fire-prone environments. We have to realize that if we value our forests, we also must value the role that fire plays in sustaining their vitality.

Literature Cited

Agee, J. K. 1993. *Fire Ecology of Pacific Northwest Forest*. Washington, DC: Island Press, 493 p.

Barrett, Stephen W. 1980. "Indians and Fire." *Western Wildlands* 6(3): 17–21.

Bosworth, D. 2003. "Fires and Forest Health: our future is at stake." *Fire Management Today* 62(2):4–11.

Carle, D. 2002. *Burning Questions: America's Fight with Nature's Fire*. Westport CT: Praeger Publishers, 298 p.

Clements, F. E. 1916. "Plant Succession: An Analysis of the Development of Vegetation." Washington, DC: Carnegie Institute Publication 242.

Foote, E. I. D., 1984. "Structure Survival on the 1990 Santa Barbara 'Paint' Fire: A Retrospective Study of Urban-wildland Interface Fire Hazard Mitigation Factors." M.S. thesis, University of California, Berkeley.

Gruell, G. E. 1985. "Indian Fires in the Interior West: A Widespread Influence." Pp. 68–74 in *Proceedings—Symposium and Workshop on Wilderness Fire: Missoula, Montana*, James E. Lotan et al., technical coordinators, November 15–18, 1983. General Technical Report INT-182. Ogden, UT: USDA Forest Service, Intermountain Forest and Range Experiment Station.

Lewis, H. T. 1983. "Why Indians Burned: Specific versus General Reasons." Pp. 75–80 in *Proceedings—Symposium and Workshop on Wilderness Fire: Missoula, Montana*, James E. Lotan et al., technical coordinators, November 15–18, 1983. General Technical Report INT-182. Ogden, UT: USDA Forest Service, Intermountain Forest and Range Experiment Station.

Martinson, E. J., and P. N. Omi. 2003. "Pre-settlement Fire Regimes of North America: A Geographic Model based on Quantitative Research Synthesis." Pp. 137–144 in *Proceedings Fire Conference 2000: The First National Congress on Fire Ecology, Prevention, and Management*, Nov. 27–Dec. 1, 2000. San Diego, CA.

Nobisso, Josephine. 2000. *John Blair and the Great Hinckley Fire*. Boston: Houghton-Mifflin Co.

Omi, P. N. 1990. "History of Wildland Burning in America from an Air Quality Perspective." Air and Waste Management Association, 83rd Annual Meeting, Pittsburgh, PA, June 24–29, 1990; reprint, 90–172.5, 7 p.

Phillips, C. B. 1985. "The Relevance of Past Indian Fires to Current Management Programs." Pp. 87–92 in *Proceedings Symposium and Workshop on Wilderness Fires*, General Technical Report INT-182, USDA Forest Service, Ogden.

Pyne. S. J. 1982. *Fire in America: A Cultural History of Wildland and Rural Fire*. Princeton: Princeton University Press, 654 p.

———. 1997. *World Fire: The Culture of Fire on Earth*. Seattle: University of Washington Press, 384 p.

Simard, A. J., D. A. Haines, R. W. Blank, and J. S. Frost. 1983. "The Mack Lake." USDA Forest Service General Technical Report NC-83, 36 p.

Steen, H. K. 1992. "The Origins and Significance of the National Forest System." Pp. 3–9 in The Origins of the National Forests, edited by H. K. Steen. Durham, NC: Forest History Society, 334 p.

Stewart, O.C. 1956. "Fire as the First Great Force Employed by Man." In *Man's Role in Changing the Face of the Earth*, W. L. Thomas, Jr. (ed.). Chicago: University of Chicago Press. Pp. 115–133.

Swetnam, T. W., and J. L. Betancourt. 1990. "Fire-Southern Oscillation Relations in the Southwestern United States." *Science* 249:1017–1020.

Williams, G. W. 1994. "References on the American Indian Use of Fire in Ecosystems." 1994. http://wings.buffalo.edu/anthropology/Documents/firebib.txt.

———. 2000. "Introduction to Aboriginal Fire Use in North America." USDA Forest Service. *Fire Management Today* 60(3): 8–12.

Wilson, C. C. 1977. "Fatal and Near-fatal Forest Fires: The Common Denominators." *International Fire Chief* 43(9): 9–15.

4

People and Events

In this chapter we examine noteworthy contributors to the study and understanding of forest and rangeland fires, including scientists, managers, and academicians. Biographical sketches and prominent events are presented, though, as in any subjective list, prominent omissions may occur. Still, the individuals noted here are considered by many to have made substantial contributions to our knowledge about wildland fires or else stand out because of their influence on the ways in which we manage forest fires today. We also explore events and legislation that have informed U.S. fire management practices.

Significant Individual Contributors

Dr. James K. Agee (b. 1945)

Fire ecologist.   Virginia and Prentice Bloedel Professor of Forest Ecology, Jim Agee teaches fire ecology courses in the College of Forest Resources, University of Washington, Seattle. He was formerly employed by the National Park Service after completing his Ph.D. at the University of California, Berkeley. His book *Fire Ecology of Pacific Northwest Forests*, published in 1993 (Island Press), is widely cited in the fire ecology literature. He also coedited *Ecosystem Management for Parks and Wilderness* (University of Washington Press).

Frank A. Albini (b. 1936)

Fire modeler.   Frank Albini is professor of mechanical engineering at Montana State University, Bozeman, Montana. He formerly worked at the USDA Forest Service Fire Laboratory in Missoula. He developed the first mainframe version of the Rothermel fire spread model during the 1970s (FIREMOD), which at the time represented a quantum leap in fire behavior prediction and the appraisal of fuelbed flammability. He has provided intellectual leadership and developed models that have greatly increased understanding of fire phenomena outside of the fire's flaming front, such as postflaming fuel consumption and maximum spot-fire distances.

Martin E. Alexander (b. 1952)

Coordinator, International Crown Fire Experiment.   Martin Alexander, Ph.D., RPF, is senior fire behavior research officer with the Canadian Forest Service, based at the Northern Forestry Center in Edmonton, Alberta, Canada. He also serves as adjunct professor with the Department of Renewable Resources, University of Alberta. Marty has been the driving force behind the International Crown Fire Experiment, an interdisciplinary effort aimed at developing improved understanding of crown fire mechanisms and impacts, based on a series of intentionally set, experimental crown fires in the boreal forests of the Northwest Territories of Canada. These experiments have provided valuable insights to a diverse variety of research concerns, including fire propagation, fire danger rating, monitoring of severe fires, durability of protective equipment, and structure ignition mechanisms.

Patricia L. Andrews (b. 1948)

BEHAVE fire prediction system.   Patricia Andrews is a research physical scientist with the USDA Forest Service Fire Sciences Laboratory in Missoula, Montana. Hired initially as a mathematician, she has been instrumental in developing and transferring fire behavior tools developed in the research laboratory to land managers and fire practitioners, for use in predicting fire behavior, planning prescribed fires, and writing fire management plans. The BEHAVE fire prediction model and its successor, BehavePlus, have been adapted for use in a variety of personal computer and hand-held applications.

Jack S. Barrows (1911–1989)

Fire researcher and education visionary.   Jack was instrumental in establishing the Northern Forest Fire Laboratory in Missoula, Montana (now known as the Intermountain Fire Sciences Laboratory of the USDA Forest Service, Rocky Mountain Research Station), one of the premier fire research facilities in the world. After graduation from Colorado State University in 1937, Jack worked as a ranger in Rocky Mountain National Park. He later studied fire behavior in the northern Rockies and hypothesized that aspen (*Populus tremuloides*) stands could function as wildfire fuelbreaks. After his retirement, he helped establish the Forest Fire Science program at Colorado State University, one of the largest in the world. This program includes an undergraduate concentration, the M.S. and Ph.D. in forest fire science, plus extensive extramural research projects spanning numerous academic disciplines.

Professor Harold H. Biswell (aka Harry the Torch) (1905–1992)

Teacher and authority on prescribed fire in California.   Known affectionately as Doc to his students, Harold Biswell was the inspiration for the use of prescribed fire in California forests from about the late 1950s. Originally from the southern United States, he was active in the early meetings of the Tall Timbers Research Center in Tallahassee, Florida, from 1962. While at the University of California, Berkeley, Doc continuously knocked heads with other faculty members and with the U.S. Forest Service over misguided fire suppression policies on public lands. Ever the diplomat and teacher, Doc officially retired from UC Berkeley in 1973, but he continued to teach classes on fire ecology at UC Davis (for two years), through university extension, and held field courses in Yosemite National Park, Mt. Diablo State Park, Calaveras State Park, and in San Diego county. His book *Prescribed Burning in California Wildland Vegetation Management*, published in 1989 (University of California Press), combined his homespun passion for fire with his personal spin on the science of fire. His legacy includes numerous students active in fire programs at academic institutions throughout the United States, as well as the reformed philosophy with which lands are managed in the West.

James K. Brown (b. 1942)

Fuel inventory, fire effects research. Jim Brown, Ph.D., pioneered techniques for quantifying fuel attributes in the field. He developed a planar intersect method for estimating fuel loadings for dead, down woody fuels by size classes, later expanded to include grasses and herbs, litter, shrubs, small trees, and fuelbed depth. These protocols are still widely followed by those seeking site-specific biomass indicators for the amount of fuel per unit area. His studies of tree crowns by fuel species provided an essential link to predicting debris and fire hazard from cutting operations. As a project leader at the Intermountain Fire Sciences Laboratory, he also performed research related to fire effects, with particular emphasis on lodgepole pine, aspen, and other plant species in the northern Rocky Mountains. He remains active even in retirement, as lead author on several key publications summarizing knowledge of fire effects on flora and on the role of coarse woody debris (CWD) in ecosystems.

George M. Byram (1909–1996)

Fire behavior researcher. George Byram made contributions in nearly every area of forest fire research. He also was a gifted painter. Perhaps he is best known for the fireline intensity and drought descriptors that still bear his name, although his biggest contributions may have been his passion for precise definitions and quantitative measures of fire behavior. He was one of the first to study fire whirlwinds, and to recognize the scaling issues with modeling of large fire behavior. His chapters 3 and 4 in the 1959 book *Forest Fire: Control and Use,* by Kenneth P. Davis, are still classics for understanding the combustion of forest fuels and fire behavior.

Craig C. Chandler (b. 1926)

Author, international textbook on fire ecology and management. Craig Chandler retired from the U.S. Forest Service as director of fire and atmospheric research after a distinguished career in fire science. His two-volume textbook, coauthored with world-renowned experts in 1982, set a standard for fire textbooks by incorporating information on fire behavior, effects, and management from an international perspective. Earlier in his career he provided insights into subjects as diverse as fire prevention and fire behavior.

A. Malcolm Gill (b. 1940)

Fire researcher, Australia.   Recently retired as senior principal research scientist at the Centre for Plant Biodiversity Research, CSIRO Plant Industry, in Canberra, Australia, Malcolm has written extensively on fire ecology, fire behavior, and fire management in a range of Australian ecosystems. He coedited *Flammable Australia*, published in 2001 (Cambridge University Press); he also was lead editor for the classic *Fire in the Australian Biota*. Throughout his career, he has focused on a wide range of fire topics, including adaptive characteristics of Australian plants to fires, such as postfire flowering, woody-fruit opening, and fire resistance in trees. He also drew together the numerous kernels of knowledge necessary for understanding the ecological effects of fires, using the fire regime concept and the adaptations of species to those regimes. His main research interests lately have centered on probability models of fire intervals (within the context of fire regimes).

Harry Thomas Gisborne (1893–1949)

Fire researcher.   Gisborne is the father of fire danger rating in the United States, a subject with a long and distinguished history. His 1928 paper "Measuring Forest-fire Danger in Northern Idaho" is one of the first writings on that complex subject, in which he attempts to integrate the effects of weather and climate on fire potential, based on his assignment at the USDA Forest Service Priest River Experiment Station. He pioneered early attempts in fire prevention and measurement of the effects of weather on fire potential through a variety of devices (some might call them contraptions), including the Asman aspiration psychrometer, visibility meter, anemohygrograph, double tripod heliograph, or blinkometer. He died in 1949 while personally investigating the site of the Mann Gulch fire fatalities in Montana (source: http://www.lib.duke.edu/forest/usfscoll/people /Gisborne/Gisborne.html).

Paul Gleason (1946–2003)

Hot-shot superintendent, fire ecologist. Fire attracts many different practitioners. Some are drawn to the excitement of facing off against a potent natural force; others are attracted by the opportunity of melding the art and science of managing fires as part of forest stewardship. Still others see in fire the beauty of mathematical truths, unfolded before their eyes as a fire makes a run up a hillside. Then again, others are drawn by the fascination with the dynamics of organized human resources united in a common enterprise—such as with fire suppression crews. Paul Gleason represented all of those different facets and much more, as a hotshot foreman, fire ecologist, mentor, and college instructor. Paul brought the same zeal and emotion to the study of fire behavior and ecology that characterized his penchant for rock climbing or understanding mathematical theorems. His passion for firefighter safety led him to develop the LCES (lookouts, communication, escape routes, and safety zones) approach to fireline construction. He was an inspiration to all whose lives he touched, both in the field and in the classroom.

Johann Goldhammer (b. 1949)

Global fire leader. Professor Dr. Johann G. Goldhammer is head of the Fire Ecology and Biomass Burning Research Group and the Global Fire Monitoring Center (GFMC) at the Max Planck Institute for Chemistry, Biogeochemistry Department, located at Freiburg University, Germany. The GFMC participates with the UN International Strategy for Disaster Reduction (UN-ISDR) and a number of international programs and provides services in global wildland fire monitoring, early warning, and strategic policy development, as well as in development of community-based fire management systems. Johann has been active in numerous international collaborations in fire management and fire science.

Edwin Vaclay Komarek, Sr. (1909–1995)

Creator, Tall Timbers Research Station and Fire Ecology conferences. Ed Komarek was the driving force behind the creation of the Tall Timbers Research Station in Tallahassee, Florida, in 1958. A visionary about the role of fire in Southern forests, he organized the Tall Timbers Fire Ecology Conference series, starting in 1962.

Initially focused on the ecological effects of fire on flora and fauna in the United States and Canada, the series has branched out to consider international fire concerns. The collection of conference proceedings (several of which are out of print but available in university libraries) represents a valuable contribution to the fire ecology literature, and a logical starting point for novices interested in the evolution of science and attitudes about fire. Over his lifetime he received numerous awards and recognitions, including several organisms named in honor of his studies. These include *Urotrema komareki* (a liver fluke), *Sigmodon h. komareki* (subspecies of cotton rat), *Komarekionidaea* and *Komarekona* (family and genus of earthworms), *Cercopeus komareki* (weevil), and *Amanita komareki* (fungus).

Norman Maclean (1902–1990)

Author, *Young Men and Fire*. Retired from the University of Chicago in 1973, Pulitzer Prize awardee Norman Maclean wrote the classic *Young Men and Fire*, completed posthumously in 1992 (University of Chicago Press) with assistance by his son John, documenting the eleven smokejumper fatalities that occurred in the Mann Gulch fire in Montana in 1949. This incident is one among several classics that come to mind when discussing historical fires and has provided inspiration for several movies, such as *Red Skies over Montana* (1952, 20th Century Fox) and *Always* (1989, Universal Studios). In 1999, Maclean's son John wrote *Fire on the Mountain: The True Story of the South Canyon Fire* (William Morrow), describing the circumstances surrounding the fourteen fire fatalities on the South Canyon fire in Colorado. John Maclean's interest in the fifteen fatalities on the Rattlesnake fire (1953) led to the publication in 2004 of *Fire and Ashes: On the Front Lines Battling Wildfires* (Henry Holt and Company).

Edward Pulaski (1868–1931)

Hero of 1910 fires and eponym for firefighting tool. Wildland firefighters are familiar with the pulaski, a practical (if not ergonomic) tool whose two-sided head cleverly merges a cutting edge (ax) with a grubbing device (hoe) for use on a fireline. This tool was the brainchild of Ed Pulaski, a ranger for the U.S. Forest Service out of Wallace, Idaho. Pulaski is remembered not only for the tool bearing his name but also for his heroics during the 1910 fires in Idaho and Montana. During the Big Blowup on August 20, 1910, Pulaski led forty-five firefighters to seek shelter from a firestorm in a narrow mineshaft near Wallace, Idaho. At one point, Pulaski reportedly stood with a pistol at the mineshaft entrance to discourage crew members from fleeing their hot, smoky shelter.

Stephen J. Pyne (b. 1949)

Author and fire historian. If there ever were a "fire laureate" designation, the title would belong to Dr. Pyne, Regents' Professor at Arizona State University West campus and author of seventeen books, thirteeen of which are devoted to the subject of fire. Worldwide, Steve is the best known contributor to the literature on fire. Publications of note (from his web page http://www .public.asu.edu/~spyne/Books.htm) include a suite of books that survey the history of fire on earth (including the United States, Europe, and Australia); see Chapter 7. He has also written many other essays and works that provide unique perspectives on firefighting, fire management, and fire science. His latest books include *Smokechasing* (2003, University of Arizona Press) and *Tending Fire: Coping with America's Wildland Fires* (2004, Island Press).

Richard C. Rothermel (b. 1929)

Fire behavior predictor. Dick Rothermel developed a set of mathematical equations for predicting fire spread (1972) based on first principles of the conservation of energy and laboratory experiments. Prior to his efforts, fire spread rates were mostly described in qualitative terms, such as low, medium, high, or extreme. His revolutionary work laid the foundation for the FIREMOD and BEHAVE fire prediction systems and subsequent computerized fire spread models, such as FARSITE, BehavePlus, and NEXUS, as well as contributing to the conceptual framework for a U.S. fire danger rating system. In 1983 he wrote a handbook, "How to Predict the Spread and Intensity of Forest and Range Fires," which has become a classic for fire managers in North America. He also authored an analysis of the Mann Gulch fire (the race that couldn't be won), which provides a sobering perspective on the topographic fire effects and implications for firefighter safety. After the Yellowstone fires of 1988, he synthesized from previous research a model for predicting the spread of crown fires over forested landscapes. Over the span of his career, Dick combined an understanding of fire behavior principles with immense respect for the tremendous challenges faced by fire practitioners. The tools he developed have aided greatly in helping firefighters do their jobs more safely.

Mark J. Schroeder (b. 1915)

Fire meteorologist.   Mark Schroeder until his retirement was a research meteorologist with the U.S. Forest Service and the National Weather Service. He became a fire weather forecaster in Chicago in the late 1940s. In 1955 he was assigned to the U.S. Forest Service as research meteorologist in fire weather and fire danger rating. At the Forest Fire Laboratory in Riverside, California (1963), he worked on many aspects of fire meteorology. Later he spearheaded development of the fire danger rating system now in use by all fire control agencies in the United States. His book *Fire Weather* (Agricultural Handbook 360, coauthored with Charles C. Buck), provides a classic primer on the role of meteorology in driving fire occurrence and behavior. He also coauthored a large study for the office of civil defense that identified surface and upper-level weather patterns and types associated with large fires in different U.S. regions. Part of the motivation for this latter study was the fear that a foreign entity might set massive forest fires as a weapon against the United States—a concern that has been expressed periodically since World War II.

Ferdinand Augustus Silcox (1882–1939)

10 A.M. policy instigator.   Gus Silcox rose through the ranks to eventually become the fifth chief, USDA Forest Service during 1933–1939. During his tenure as chief, the Forest Service instituted its infamous 10 o'clock policy, calling for the use of any means necessary to control an ongoing wildfire by 10 A.M. the next morning. Detractors point to this policy as creating the context for costly fire exclusion policies that created unhealthy forest conditions throughout the western United States by the end of the twentieth century. In fairness, Chief Silcox was at the helm during the Great Depression, when the Civilian Conservation Corps and Works Projects Administration effectively helped millions of unemployed workers find gainful employment working on construction projects in the national forests. The massive workforce available may have prompted government administrators to believe that fires could actually be controlled with such dispatch—with relatively little need to consider ecological impacts at that time.

Herbert Stoddard (1889–1970)

Early research pioneer, bobwhite quail habitat.   Herbert Stoddard was an early pioneer in the use of fire in longleaf pine ecosystems of the southern United States. An outdoorsman and self-taught ecologist, forester, and naturalist, he was an early proponent for using fire as a management tool for use in improving bobwhite quail (*Colinus virginianus*) habitat. In 1941 he formed a forestry consulting business in Georgia to advise private landowners on managing longleaf pine forests, a continuation of quail research he had started earlier and for which he had recruited the services of Ed Komarek. Both recognized that fire exclusionary policies were leading to declines in local bird populations. He provided the inspirational vision for conception of the Tall Timbers Research Station, many years before its eventual creation in 1958.

Charles E. Van Wagner (b. 1924)

Canadian fire specialist.   Charlie van Wagner helped pioneer fire behavior research in Canada. His writings on crown fires and crown scorch, even if based on limited samples, were considered robust enough to be incorporated in recent U.S. fire behavior and fire effects models. He was one of the first to write about the divergent paths taken by North American fire behavior researchers, with the United States adopting a laboratory model relying on first principles of the conservation of energy, versus the Canadian empirical approach. Since 1925, Canada has had a rich history of fire research that parallels developments in the United States (conducted by the likes of Herbert W. Beall and Jim Wright).

Domingos Xavier Viegas (b. 1950)

Coordinator, International Forest Fire Conferences.   Professor Viegas, of the Mechanical Engineering Department at the University of Coimbra in Portugal, has coordinated numerous trans-Europe research projects and hosted several international fire conferences, focusing on the latest scientific findings from around the world. He also has written numerous articles on fire spread and fire management that have helped to elevate the profile of fire science among members of the European community.

Historic Events

A brief historical overview of forest management in the United States provides a useful context for understanding the evolution of fire management. The history of fire management in the United States is rooted in forest policies, large fire events, legislative precedents, influential personalities involved with management of federal public lands, and more. In many ways, U.S. fire history mirrors the structural and philosophical evolution of the agencies that have managed fires on our behalf for almost a century. These agencies in turn reflect the dominant social mores and political undercurrents at work within both developed communities and wildland ecosystems, as affected by science and human values.

In Chapter 3 we chronicled the eras characterized by large fires. Here we describe the legislative acts and significant events that accompanied some of those large fires. Several of the same fires are included in Tables 4–1 through 4–6, for cross-referencing with their descriptions in Chapter 3.

Frontier Settlement Era (Table 4–1)

During the early twentieth century, settlers saw little organized response to uncontrolled wildfires in forests and rangelands. Many fires devastated forests and communities, such as the 1871 Peshtigo fire, which burned more than 3.5 million acres (1.4 million ha) in Wisconsin and Michigan, leaving 1,500 dead. In 1902, the Yacolt fire in southwestern Washington burned about a million acres (405,000 ha) and cost thirty-eight lives. During that period, Forest Service personnel and area residents fought fires primarily on federal, state, and private lands. It was mostly hand-to-hand combat with wildfire, using crude tools such as wet burlap bags, axes, and water buckets.

The first forest reserve (Yellowstone National Park Timberland Reserve of Wyoming) was established in March 1891 by proclamation following President Benjamin Harrison's signing of the so-called Forest Reserve Act of 1891 earlier that same month.

In October 1891, President Harrison proclaimed the creation of the second reserve, the White River Plateau and Timberland Reserve of Colorado (Lynch and Larrabee 1992).

Creation of the forest reserves followed on the heels of a century of perceived timber exploitation, wildlife slaughter, overgrazing of grasslands, and rampant forest fires (Vance and Vance n.d.). Reserves were to be protected from exploitation and managed by the federal government in behalf of the national interest. Actually, then as now, controversies reigned over the actual status of natural resources on public lands, with detractors proposing theories of pending or actual timber famines and other resource despoliation. Pitted against the doomsayers were supposed visionaries and field practitioners who believed that the salvation of the considerable natural resource wealth of the United States could be found in wise use and management. This battleground of viewpoints persists to the present regarding the status and options for managing public forest resources.

The first forest reserves were placed under the responsibility of the Department of Interior's General Land Office. At that time, federal foresters worked for the Bureau of Forestry in the Department of Agriculture. By 1892, fifteen reserves had been created, encompassing 13 million acres. Most were created to protect the

Table 4-1
Applicable Events, Pivotal Large Fires, and Laws Governing Federal Fire Management during the Frontier Settlement Era

Year	Event/Fire/Law
1871	Peshtigo Fire
1872	Yellowstone Park created
1891	Creative Act of 1891 (26 Stat. 1103, 16 USC 471), also known as the Forest Reserve Act, allows President to set apart reserves from the public domain for purposes such as watershed, timberland, or wildlife protection.
1897	Organic Act of 1897 (30 Stat. 35) creates national forests to protect water flows and timber supply. Authorizes Secretary of Interior to make provisions for the protection from fire and other depredations.
1905	Transfer Act of 1905 (33 Stat. 628, USC 472) gave responsibility for administration of national forests to the Department of Agriculture (from Interior).
1908	Emergency Firefighting Fund allowed the federal government to go into debt to fight forest fires.

Sources: Pyne 1982; San Juan National Forest 1993.

headwaters of important streams at the behest of irrigators and real estate developers (Pisani 1992).

Early federal foresters were expected to participate in fire suppression activities as one of their primary duties. However, the purpose and administration of the forest reserves weren't clarified until passage of the 1897 Organic Act, which provided for fire protection among the other supervisory and administrative duties needed to manage the reserves. In 1905, President Teddy Roosevelt transferred the forest reserves to the jurisdiction of the Department of Agriculture, in the Bureau of Forestry, at that time renamed the Forest Service. The forest reserves were renamed national forests in 1907 (Steen 1992), today encompassing 191 million acres, many of which support flammable ecosystems.

Early Years of Forest and Fire Management (Table 4–2)

Wildfires and their control dominated the agenda of the youthful federal agencies at the turn of the twentieth century, although the crude tools available to the understaffed crews were no match for large fires. As such fires occurred across the United States, state and local governments became more involved and concerned about costs and resource damage. After the severe wildfires (that is, the "Big Blowup") in Montana and Idaho in 1910, more emphasis could be placed on newer techniques, such as telephone communications and fire patrols. State fire warden positions were established, especially in the West, and legislation such as the Weeks Act in 1911 and the Clarke-McNary Act in 1924 enhanced the federal-state fire suppression partnership. Those programs grew into the U.S. Forest Service's State and Private Forestry Division, of which Fire and Aviation Management is a part today. In 1944, Congress increased the scope of the Clarke-McNary Act to create an emphasis on fire prevention. (source: http://www.fs .fed.us/fire/people/aboutus.html).

The great Idaho-Montana fires of 1910 attracted great attention, but elsewhere (such as in northern California and the South) the use of low-intensity surface fires was being proposed as an alternative to all-out fire suppression. Advocates of "light burning" included a loose amalgam of scientists, magazine writers, and private timber company owners who recognized that periodic surface fires would help to keep fuel hazards more manageable.

Forest burning by Native Americans was recognized as a power-ful influence on ecosystem structure that had been removed by Euro-American settlement (see Chapter 3). At the same time, the U.S. Forest Service viewed light burning as a threat to its mission to protect public forests from fire. In essence, the agency felt that the public would not be able to digest a mixed message about fire's destructiveness along with its potential benefits.

The National Park Service was created in 1916 to promote and regulate the use of federal national parks, monuments, and reser-vations. The enabling act provided for the secretary of interior to appoint Stephen T. Mather as the first NPS director (with a salary of $4,500 per year) to oversee the system of national parks, "whose purpose (was) to conserve the scenery and the natural and historic objects and the wildlife therein and to provide for the enjoyment of the same in such manner and by such means as will leave them unimpaired for the enjoyment of future generations" (source: http://www.nps.gov/legacy/organic-act.htm). From sixteen parks in 1916, the NPS has grown to 388 units today, encompassing units as diverse as national parks, monuments, preserves, historic sites, historical parks, memorials, battlefields, cemeteries, recreation areas, seashores, lakeshores, rivers, park-ways, and trails.

Table 4-2

Applicable Events, Pivotal Large Fires, and Laws Governing Federal Fire Management during the Early Years of Forest and Fire Management in the United States

Year	Event/Fire/Law
1910	Great Idaho Montana Fires burned over 3 million acres.
1911	Weeks Act (36 Stat. 961) The act allowed for 1) the purchase of land at the head of navigable streams as national forest; 2) cooperative agreements and matching funds between the USFS and state foresters for fire protection. (source: Pyne 1982, p. 61).
1916	National Park Service Organic Act of 1916 (39 Stat. 535 plus amendments) created the National Park Service.
1921	US Forest Service Chief Greeley convenes the agency's first national conference, the Mather Field Conference of 1921.
1924	Clarke-McNary Act (43 Stat. 653-654), as amended; 16 U.S.C. 564, 565, 566, 567) expanded provisions of the Weeks Act to include all watersheds, not just navigable waterways. (source: Pyne 1982) Enabled federal assistance on state and private lands.

Sources: Pyne 1982; San Juan National Forest 1993.

The U.S. Forest Service held its first national conference, the Mather Field Conference of 1921, essentially setting the tone for ending the debate about light burning in favor of systematic fire protection of the nation's forests (Pyne 1982). The proposed "light burning" generated considerable controversy then, and it remains controversial to the present day.

Institutionalization of Fire Exclusion (Table 4–3)

The Great Depression saw a continuation of large fires in the United States, such as the Tillamook fire in western Oregon, though the nation's poor economic conditions created a vast unemployed labor pool that could be put to work in the forests. In 1935 the Forest Service adopted its now infamous "10 A.M. policy," an extension of the idea that fires could be systematically suppressed across the country. Accordingly, the policy stipulated that a fire was to be contained and controlled by 10 A.M. following the report of a fire, or, failing that, by 10 A.M. the next day— and so on. This policy thus provided rationale for the use of all fire control resources required to suppress a fire as soon as possible—in the process, justifying large expenditures as well. The policy remains in effect today in certain areas where property in harm's way requires aggressive protection, although the policy has been largely rejected for some forest types—such as ponderosa pine ecosystems in the southwestern United States.

Firefighting took to the air shortly after World War I as the Forest Service used aircraft to patrol for wildfires. Smokejumpers came onto the scene to fight their first fire in 1940. In 1956 the first practical aerial delivery of water and chemicals onto wildfires began, and helicopters began to assist with firefighting in the 1950s (source: http://www.fs.fed.us/fire/people/aboutus.html). Today, airplanes are the most expensive means employed to fight wildfires.

During World War II, firefighting was considered a patriotic act, tantamount to waging war with the enemies of the United States. Keep Green movements were spawned in the Pacific Northwest to assist the U.S. government in wildfire prevention efforts. Smokey Bear became a national symbol for rallying the country against wasteful fires by preventing them from occurring in the first place. The Federal Property and Administrative Services Act of 1949 permitted the transfer of excess property (except

Table 4-3
Applicable Events, Large Fires, and Laws Governing Federal Fire Management during the Time Period in which the Fire Exclusion Policy Was Institutionalized in the United States

Year	Event/Law/Fire
1933	During August, 1933, the Tillamook Fire burned across 311,000 acres in Oregon. 1 life was lost; some of the same areas burned again in 1939.
1935	US Forest Service Chief Silcox instituted the 10 AM Policy.
1940	Keep Oregon Green Association was founded.
1944	The US Forest Service in conjunction with the Advertising Council originated and authorized a poster by Albert Staehle of Smokey Bear as the symbol for fire prevention.
1945	Keep Montana Green Association was founded.
1946	Keep Idaho Green Association was founded.

Sources: Pyne 1982; San Juan National Forest 1993.

real estate) among federal agencies and to other entities, eventually enabling fire agencies to access property such as aircraft and trucks that could be customized for firefighting.

Policy Reflections (Table 4–4)

The Cold War with the USSR and fears of communist conspiracies brought along concerns that multiple fires set in the backcountry could pose a serious threat to U.S. internal security. Hysteria about wildfires as a possible weapon was heightened as the Bel Air fire burned into suburban Los Angeles in 1956. At the same time, the U.S. Forest Service sought to recodify its doctrines of wise use of multiple resources, even as the National Park Service was reconsidering its management philosophy as a result of the 1963 Leopold Report findings (see Chapter 3).

Passage of the 1964 Wilderness Act and the 1969 National Environmental Policy Act laid important groundwork for positions that would be adopted by environmentalists to ensure that management activities were properly conceived and implemented. The 1967 Clean Air Act (and subsequent amendments) likewise would provide a regulatory context for ensuring that air quality would not be harmed by federal land management activities within individual states. Ironically, prescribed fires intended to stave off wildfire hazards are subject to emission controls, whereas wildfires are considered a natural hazard and not subject to regulation.

Table 4-4
**Applicable Events, Large Fires, and Laws Governing Federal Fire
Management during 1950–1969**

Year	Event/Fire/Law
1950	A bear cub discovered after a wildfire in NM is nicknamed Smokey Bear and adopted as a national symbol for fire prevention.
1956	Bel-Air fire burns over 500 homes in southern California.
1960	Multiple Use-Sustained Yield Act of 1960 (16 USC 528) authorizes and directs Secretary of Agriculture to develop and administer renewable surface resources of national forests; authorizes cooperation with state and local governments.
1963	Leopold et al. (1963) Report calls for national parks to represent a vignette of primitive America, including habitats manipulated by prescribed fire.
	Federal Clean Air Act (PL 88-206) empowers the Secretary of Health, Education, and Welfare to define air quality criteria based on scientific studies.
1964	Wilderness Act of 1964 (16USC 1131-1136 provides for establishment and administration of the national wilderness preservation system.
1967	Federal Air Quality Act (PL 90-148) establishes a framework for defining "air quality control regions" based on meteorological and topographical factors of air pollution.
1969	National Environmental Policy Act of 1969 (42USC 4321-4335) requires disclosure of the environmental consequences associated with implementation of a proposed action and alternatives.

Sources: Pyne 1982; San Juan National Forest 1993.

Environmentalism Takes Off (Table 4–5)

The environmental legislation of the 1960s spawned an outpouring of related laws and agency mandates in the 1970s as well. The 1970 Clean Air Act is the comprehensive federal law that regulates air emissions (including smoke) from area, stationary, and mobile sources. This law authorizes the U.S. Environmental Protection Agency to establish National Ambient Air Quality Standards (NAAQS) to protect public health and the environment. The goal of the act was to set and achieve NAAQS in every state by 1975. The setting of maximum pollutant standards was coupled with directions that states should develop state implementation plans (SIPs) applicable to appropriate industrial sources in the state. The act was amended in 1977 primarily to set new goals (dates) for achieving attainment of NAAQS, since many areas of the country had failed to meet the deadlines. The 1990 amendments to the Clean Air Act in large part were intended to meet unaddressed or insufficiently addressed problems such as acid rain, ground-level ozone, stratospheric ozone depletion, and air toxins.

The 10 A.M. policy for fire control (see Table 4–3) was modified slightly in 1971 when the U.S. Forest Service decided to aim its presuppression efforts (for example, fuel treatments) at curtailing fires of 10 acres (4 hectares) or less. The so-called 10-acre policy for presuppression planning thus represented formal recognition of the importance of managing fuels in conjunction with overall preparedness planning. Equally significant, the 10-acre planning standard represented a first formalized departure from the agency's 10 A.M. policy initiated in 1935. Even so, the 10 A.M. policy remained the de facto policy for managing fire in wildland areas until recently. In most areas, wildfires are attacked as soon as practicable in order to keep the fire as small as possible and to reduce damage to valued resources. Ironically, fuels in some of these areas may have been regulated historically by recurrent, low-severity surface fires. Prescribed fire may be a viable option for managing fuels in these areas, especially when ignited under conditions that mimic past fires.

The enhanced federal commitment to fuels management in the 1970s was stimulated in part by calamitous fire seasons in southern California and the Pacific Northwest in 1970–1971. Massive infusions of funds for fire presuppression early in the 1970s were later derailed by congressional concerns over the lack of visible returns for fuel treatment investments.

Congressional acts, such as the 1974 Resources Planning Act and the 1976 National Forest Management Act, mandated forest and fire planning as well as consideration of the environmental impacts of management decisions. The 1976 Federal Land Management and Policy Act provided similar direction to the Bureau of Land Management.

In 1978, the U.S. Forest Service revised its fire policy to call for cost-effective fire management and reliance on fire use strategies, primarily prescribed fire. From the 1970s onward, the economics of fire management became increasingly important as Congress, the Government Accounting Office, the Office of Management and Budget, and others expressed growing concerns over the high cost of firefighting.

The 1980 Mack Lake fire (see also Table 3–5) nearly resulted in permanent termination of all prescribed fire programs on national forests, so great was its perceived harm. The houses destroyed and fatality were bad enough, but more critically, the fire

Table 4-5
**Applicable Events, Large Fires, and Laws Governing Federal Fire
Management during the Period of Increased Environmentalism**

Year	Event/Fire/Law
1970	Clean Air Act of 1970 (42 U.S.C. s/s 7401 et seq.) (1970)
1971	USFS adopts a 10-ac standard for pre-suppression planning; Environmental Protection Agency (EPA) promulgates national ambient air quality standards for particulates, photochemical oxidants (including ozone), hydrocarbons, carbon monoxide, nitrogen dioxide, and sulfur dioxide.
1974	Forest and Rangeland Renewable Resources Planning Act of 1974, as amended by the National Forest Management Act of 1976 (16 USC 1600-1614) institutes land and resource management planning and planning regulations for the US Forest Service. Forest plans must comply with Acts of Congress enacted to protect air, water, threatened or endangered species, cultural, or historic resources.
1976	Federal Land Policy and Management of 1976 (90 Stat. 2743) mandates planning for lands administered by Bureau of Land Management; original Smokey Bear dies at the National Zoo in Washington DC. (Burial at Capitan NM)
1977	Revised Fire Policy of USFS calls for cost-effective fire management, including fire use; Clean Air Act Amendments (PL 95-95) set the goal for visibility protection and improvement in Class I areas and assigns federal land managers with the affirmative responsibilities to protect values related to air quality.
1984	Rudolph Wendelin designed a 20-cent postage stamp depicting a bear cub clinging to a burnt tree with the famous Smokey Bear emblem as a background. This represented the first and only time the U.S. Postal Service issued a postage stamp honoring an individual animal.
1988	Greater Yellowstone fires burned nearly half of the national park.
1989	Federal Policy re-evaluated after Greater Yellowstone fires.
1990	Clean Air Act Amendments (PL 101-549) establish authority for regulating regional haze and acknowledge the complexity of the relation between prescribed and wildland fires.

Sources: Pyne 1982; San Juan National Forest 1993.

had been set intentionally by U.S. Forest Service personnel as a prescribed burn that then escaped control.

Fatalities and Houses Take Center Stage (Table 4–6)

Recent fire episodes—for example, Florida (1998), Colorado (2002, 1996, 1994, 1990, and 1989), Oregon and Arizona (2002),

Washington and Idaho (1994, 1991), Montana (1994), southern California (2003, 1996, and 1993), Oakland-Berkeley (1991), and Yellowstone (1988)—illustrate problems of increasing severity, complexity, and cost. These fires have destroyed life and property, involved multiple agency and private jurisdictions, and cost billions in suppression expenditures and resource damage. In addition, these fires have provided evidence that land uses and management practices (that is, fire exclusion) during the twentieth century have unintentionally increased fuel hazards in some areas.

Severe and tragic fire seasons have begotten numerous reviews of fire policy and operational procedures, as agencies try to learn from large fires and to cope with their effects. The fourteen fatalities in Colorado in 1994 spawned the 1995 Federal Fire Policy review. The 2000 fire season resulted in the National Fire Plan (during the administration of President Clinton) for restoring fire to Western forests, especially those from which fire had been excluded during the twentieth century (see Chapter 6). In the midst of the 2002 fire season, President Bush proposed his Healthy Forests Initiative to accelerate thinning projects in Western forests, including what some perceived as limitations on environmental scrutiny and judicial review.

The Healthy Forests Restoration Act, passed by both houses of Congress and signed by President Bush in 2003, provides for the following (USDA Forest Service 2003):

- Strengthening public participation in developing high-priority forest health projects;
- Reducing the complexity of environmental analysis;
- Providing a more effective appeals process encouraging early public participation in project planning;
- Instructing courts when asked to halt projects to balance the short-term effects of project implementation against the harm from undue delay and the long-term benefits of a restored forest.

A critical part of the initiative builds on congressional legislation enacted in 2002 allowing federal agencies to enter into long-term stewardship contracts (of up to ten years) with small businesses, communities, and nonprofits. Accordingly, these entities can use the proceeds from tree thinning and removal to defray the costs of hazard reduction operations as well as to derive profits from cut-

Table 4-6
Applicable Events, Large Fires, and Laws Governing Federal Fire Management since 1990

Year	Event/Fire/Law
1994	South Canyon Fire—14 firefighter fatalities (35 nationwide)
1995	Federal policy review after the South Canyon fire result in changes to firefighter and public safety priorities; Joint Fire Science Program initiated several years later.
1999	Environmental Protection Agency (EPA) promulgates regional haze rules.
2000	2000 fire season worst in 50 years; National Fire Plan of the Clinton administration.
2001	Federal prescribed fire re-evaluated after Cerro Grande escaped prescribed fire near Los Alamos, NM.
2002	Largest fires on record occur in Colorado, Arizona, and Oregon.
2003	Healthy Forests Restoration Act; Southern California fire complex is the most expensive in state history.

Sources: Pyne 1982; San Juan National Forest 1993.

ting operations. Environmentalists have expressed fears that the initiative will result in excessive tree removals, particularly in the larger diameter trees that increase timber sale revenues, and will increase road building in remote forest areas.

Historical Events Summary

As public fire management agencies have evolved, so have their policies—often in the midst of considerable controversy. Early fears of timber famine, floods, and droughts have in some sense given way to environmental disputes over appropriate uses of the nation's public forests. However, then as now, the basic dispute seems to pit those who believe that the forests have been abused and overused against those who prefer to see no logging or roads constructed in the woods. Fuels management, whether by mechanical thinning, prescribed fire, or other means, has spawned its share of debate between those who view thinning as a hazard reduction tool and those who suspect that thinning provides an excuse for the logging of larger trees on public land.

For most of the twentieth century, federal fuels management was mostly synonymous with wildfire hazard reduction. Fuel treatments, such as prescribed fire and mechanical thinning, were aimed largely at reducing the size and severity of eventual wildfires in the treated area. The passage of environmental legislation in the 1960s and 1970s broadened considerably the concerns associated with proposed fuel treatments, especially if and where endangered species, cultural resources, or environmental influences (that is, air, soil, and water) might be affected. In such instances, justification for the proposed treatment must survive scrutiny regarding its effects on the variety of resources in the area as well as downwind and downstream.

The U.S. Forest Service (2003) estimates that 190 million acres (77 million hectares) of public lands are at elevated risk of severe wildfires, yet the magnitude of the problems and potential pitfalls are poorly documented. For example, information is lacking on the extent to which wildfire threats can be reduced through the projected increases in fuel treatments. We don't know the optimal sizes and timing of areas to be treated (by prescribed fire or other fire surrogates) in order to reduce wildfire costs and losses successfully. With the current anticipated backlog of areas in need of treatment, there is no guarantee that the projected increases will mitigate future wildfire outbreaks. Furthermore, treatment responses will vary by fire regime, ecosystems affected, and treatment method. Some fire regimes may not require hazard reduction treatments, as fuels have not accumulated abnormally despite fire suppression efforts. Finally, in the absence of maintenance efforts such as fire or mechanical thinning/removal, the effectiveness of treatment may decline over time, so that chance ignitions may become wildfires even in treated landscapes. These examples illustrate just a few of the problems associated with incomplete understanding of hazardous fuel problems in the nation's forests and rangelands.

These and other information gaps could lead the U.S. Congress and other policy makers to develop unreasonable expectations for the possible accomplishments for the interagency fire science and management initiatives. For example, extreme wildfire years could occur while treatments are in progress or even shortly after completion, depending on location and timing of chance ignitions and climatic conditions, such as extreme drought. Fires could occur in or adjacent to treated areas, depending on the size and effectiveness of fuel modification. Expansions in prescribed fire programs could meet with resistance in the absence of better information on the necessity and effects of fuel treatment in an area, just as the program was sidetracked during the 1970s.

These considerations are especially important in ecosystems of the western and southeastern United States that are susceptible to large fire outbreaks on account of prolonged drought, hazardous fuels, and lightning strikes. In the Southeast and West, fuels treatment (for example, prescribed fire plus other cultural, mechanical, biological, and chemical methods) has been practiced for years, but knowledge is not widely available about the effects on actual hazard reduction and other biological, social, and economic consequences. In addition, little synthesis has been attempted across ecosystems, agency jurisdictions, and land-use patterns. Several studies have attempted to integrate findings across agencies and ecosystems, but much work remains.

Today fire is recognized as an important part of healthy, diverse ecosystems, but a force that usually must be suppressed for reasons of public safety. The policy of fire suppression on every fire (adopted in the early twentieth century) has been reformed to focus on fire suppression where needed, with fire use (prescribed fire and natural ignitions) elsewhere to achieve resource objectives as appropriate. However, applying this dual perspective toward fire without error on public forests is more easily said than done, and probably impossible given the total area of lands to be managed by natural resource agencies. Even so, fires and their management have provided high visibility to agencies with fire responsibilities on public and private lands, with both amazing success stories and wrenching failures. As the image of fire has undergone numerous transformations over the past century, back and forth between hated enemy and useful friend, the agencies have had to contend with a public that seems unsure of what lies ahead. Regardless, the story promises to continue as fascinating and gripping as ever.

Literature Cited

Leopold, A. S., S. A. Cain, C. M. Cottam, I. N. Gabrielson, and T. L. Kimball. 1963. "Study of Wildlife Problems in National Parks." Pp. 28–45 in Transactions 28th Conference on North American Wildlife and Natural Resources.

Lynch, D. L., and S. Larrabee. 1992. "Private Lands within National Forests." Pp. 198–216 in *The Origins of the National Forests,* edited by H. K. Steen. Durham, NC: Forest History Society, 334 p.

Pisani, D. J. 1992. "Forests and Reclamation, 1891–1911." Pp. 237–258 in *The Origins of the National Forests,* edited by H. K. Steen. Durham, NC: Forest History Society, 334 p.

Pyne, S. J. 1982. *Fire in America.* Princeton: Princeton University Press, 654 p.

San Juan National Forest. 1993. "Amended Land and Resource Management Plan." USDA Forest Service San Juan National Forest, Durango, CO.

Steen, H. K. 1992. "The Origins and Significance of the National Forest System." Pp. 3–9 in *The Origins of the National Forests,* edited by H. K. Steen. Durham, NC: Forest History Society, 334 p.

USDA Forest Service. 2003. "Implementation of the Healthy Forests Initiative." Release No. fs0405.03. Available at http://www.usda.gov/news/releases/2003/12/fs0405.htm.

Vance, M. R., and J. A. Vance. N.d. "The Story behind the Pike National Forest." http://www.fs.fed.us/r2/psicc/pp/history.htm.

5

Facts and Data

Fire Activity

Worldwide, an area larger than half the size of mainland China may burn annually, with about 90 percent of the ignitions caused by humans (Nature Conservancy 2003). This translates into about 1.9 million square miles (4.9 million square kilometers) burned globally every year. Around the world, fires occur in temperate ecosystems, boreal and tropical forests, and savannas and grasslands—essentially wherever and whenever flammable biomass and ignition sources interact to support combustion. In any given year, wildfires may burn on all the world's continents, except Antarctica. By comparison, about 100,000 fires burn about 2 to 8 million acres (0.81–3.24 million ha) annually in the United States. Most of these fires receive scant attention because of their limited extent, remoteness, and apparent lack of impact. In fact, most fires burn with little fanfare and go out or are controlled by firefighters with effects localized to the immediate area of the burn. However, relatively few fires (that is, less than 3 to 5 percent of the total number of ignitions in the United States) may result in most of the burned area, costs, and losses. Globally, the impacts of burning on greenhouse gases, climate change, and other atmospheric processes are poorly understood, although of growing concern. In this chapter we examine some of the facts and data about fires that are well understood, including summaries of fire occurrence, combustion as physical and chemical processes, interactions between fire and the environment, fuels and fire behavior prediction, fire effects, and societal responses in terms of management activities. We also look at employment opportunities in forest fire science across North America.

Major fire activity is episodic over time (Figure 5–1), with peaks and dips in fire occurrence and the area burned. Generally, more damage and concerns are associated with the peak fire years, although costly fires occur in all years. For example, 125,000 fires burned more than 8 million acres (3.24 million ha) in 2000, considered the most damaging fire season in nearly half a century. In most years, relatively few fires account for most of the burned area. In some years—for example, 1998—relatively few fires (roughly 70,000) burn a small total area (less than 3 million acres, or 1.2 million ha). In 1975, twice as many fires (some 140,000) burned less than 2 million acres (0.81 million ha). In fact, the 1970s had a greater number of fire starts and relatively lower area burned than the 1990s, when fewer ignitions were associated with several peaks in burned area.

The usual causes for the variability in burned areas over time include climate and fuels. In some years, the cumulative effects of ongoing drought result in desiccated fuels early in spring—often a harbinger of massive summer burns if no rains fall and ignitions

Figure 5-1
U.S. Fire Frequency and Area Burned, 1960–2000
Annual U.S. ignitions and area burned vary considerably from year to year.

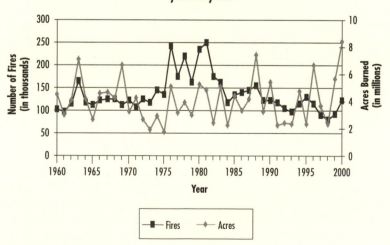

Source: http://www.nifc.gov/stats.
Note: 1 ac =.405 ha

occur during periods of high wind or prolonged heat. In reality every fire season represents a cumulative expression of many influences, including the antecedent weather, vegetation dynamics (plant growth and succession), and a variety of human activities. A relatively wet year with few fires may beget an extreme fire season subsequently, as the excess moisture produces a lush plant cover that becomes combustible when dried.

The relationship between fire frequency and area burned changes constantly (Figure 5–1). In fact, the factors that explain forest fire ignitions and burn patterns will change from year to year, depending on a variety of influences—namely, weather and climate, fuels, and human activities, to name a few. Other factors that affect fire activity in an area include the onset of drought (that is, El Niño/La Niña cycles), location of lightning storm tracks and topographic position, and proximity of human population centers. Figure 5–2 shows even greater fluctuations if we go further back in time and examine the number of fires and average area burned by decade. From a decadal perspective, the average annual number of fires is fairly stable (roughly 100,000 to 150,000), although area burned has dropped dramatically since the 1930s (about 40 million acres burned per year back then) to less than 5 million annual acres during the 1990s. In fact, the average fire size (calculated by dividing the area burned by the number of fires) decreased from 234 acres in the 1930s decade to 34 acres in the 1990s. Actually, since the 1930s the average size of fires steadily dropped to a low of 21 acres in the 1970s and has increased slightly since then to 34 acres in the 1990s. The drop-off in area burned is most likely the result of technological improvements in the way fires are fought, as well as the institutionalization of fire suppression by public land management agencies. The increase since the 1970s may be due to climatic warming, as well as agency policies that permit fires to burn larger areas under certain prescribed circumstances. Some attribute the increase in area burned to fuel accumulations resulting from twentieth-century fire exclusion policies, although that assertion is difficult to prove systematically for all ecosystems.

The dramatic drop-off in average area burned since the 1930s (Figure 5–2) is a tribute to the effectiveness of fire suppression policy and technology, but it also gives a crude idea of the amount of fire that has been withheld from the national landscape during the same time period. Although this is speculation, it is safe to say that some, though certainly not all, of the fires that have been sup-

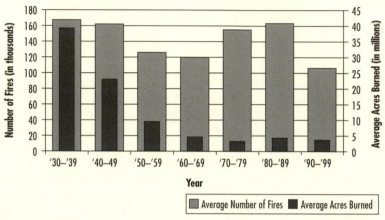

Figure 5-2
Average Number of Fires and Area Burned in the United States by Decade Since 1930
The Average area burned has dropped dramatically since the 1930s, though episodic years persist.

Source: http://www.nifc.gov/stats.
Note: 1 ac = .405 ha

pressed over the years might have performed important ecological functions had they been allowed to burn. If allowed to burn, those same fires might have maintained tree densities and fuel loads at sustainable levels, while contributing to biological diversity and healthy ecosystems. Obviously we don't want to turn back the clock to 1930, or even try to imagine how a manager might attempt to select the time and place for selecting and shepherding just the "beneficial" ignitions while suppressing all others. But it is precisely that balance between wanted and unwanted fires that is needed when managing the suppression and use of fire over a large area.

Of course, decadal trends (for example, Figure 5–2) may be a little misleading, since fires do not necessarily behave differently just because of the passage of time. Also, averages depicted in Figure 5–2 mask annual variability, although trends can emerge by looking at fire occurrence and area burned at different time scales. For example, climatic dryness in the Pacific Northwest, known as the Pacific Decadal Oscillation, has been suggested as the possible driving force behind fire cycles that run on a roughly ten-year basis in the northern Intermountain West (Zhang et al.

1997). So analyses over multiyear time frames are instructive because they can suggest important trends that may be overlooked in a year-to-year analysis. Furthermore, we need to remember that forest fire trends may be driven by a multitude of causes, including climatic as well as social drivers, some of which are either poorly understood or perhaps have yet to be discovered.

Let's look at another ten-year period to examine trends in U.S. human- versus lightning-caused fires (Figure 5–3). For the ten-year period 1988–1997, an average of 116,000 fires burned 4 million acres per year. During that ten-year period, in any given year humans accounted for nearly 103,000 fire starts (about 87 percent of the total), burning about 1.9 million acres (770,000 ha) per year. In the same period, lightning accounted for about 13,000 fires, which burned about 2.1 million acres (851,000 ha) annually. The proportion of fires caused by humans or lightning will vary by region. In the mountainous West or in the Southwest, lightning may start 80 to 90 percent of the fires, thus reversing the proportions observed nationwide. By contrast, in southern California, 95 percent of ignitions may be due to human causes. Thus humans are the primary fire-starters in more populated areas, while lightning may predominate in isolated, mountainous areas.

Lightning provides a spectacular ignition source as thunderbolts crash to the ground. Yet the mechanisms that produce lightning and lightning fires are poorly understood. Lightning literally surrounds our world, with as many as 44,000 thunderstorms and 8 million lightning flashes taking place daily. A typical lightning fire in the forest originates from electrical currents discharged between the cloud and ground, with high temperatures in the vicinity of the lightning strike apparently raising fuel temperature to the point of ignition. The likelihood of a lightning ignition producing a forest fire depends on fuel moisture and the type and duration of discharge, among many factors. In the United States, lightning fires are most prominent in the northern and southern Rockies, the Cascades, and the Sierra Nevada mountain ranges, where hundreds of fire starts may result from the passage of a single storm system. More thunderstorms occur in the Southeast, but they start fewer fires because of the accompanying rain (Schroeder and Buck 1970).

The data that were used to develop Figure 5–3 show that the average size of lightning fires tends to be larger than that of human-caused fires—152 acres (62 ha) versus 19 acres (8 ha) during 1988–1997)—since lightning tends to occur in more remote

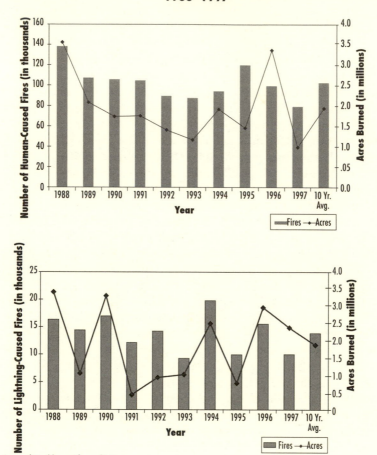

Figure 5-3
Number and Size of Wildland Fires in the United States,
1988–1997

Source: http://www.nifc.gov/stats
Note: 1 ac = .405 ha.

areas where ignitions may be more difficult to detect. Such areas also may be less accessible to initial attack crews. Thus more time may elapse, and the fire will continue to grow, before fire control forces arrive to initiate suppression activities. Also, a small proportion of lightning ignitions in park and wilderness areas may not be fought aggressively, but instead may be allowed to burn into natural barriers (such as rocks or lakes) or be constrained in size only by low-impact herding strategies until extinguished by

the autumn moisture. Such strategies are employed to allow a
more natural role for fire in designated areas. Specific management plans are required if agencies are to allow natural ignitions
to spread under these conditions, but once the planning is complete, some fires may burn for weeks or even months at a time,
growing until extinguished by rain or snow at the end of the fire
season. Although historically restricted to relatively few areas,
this practice is expanding as managers and the public accept a
more natural role for fire. These so-called "wildland fire use" incidents (formerly called prescribed natural fires) will likely increase in frequency and size in the future, as public agencies develop greater competency and familiarity with strategies for
herding ignitions to achieve their resource management objectives. Although the impact of lightning fires may not be so
strongly apparent in the data presented, their average size during
1988–1997 ranged from a low of 40 acres (16 ha) in 1991 to a high
of 267 acres (108 ha) in 1997. By contrast, during the same period
the average size of human-caused fires ranged from a low of 12
acres (5 ha) in 1995 and 1997 to a high of 34 acres (14 ha) in 1996.

Suffice to say that the number and size of fires vary considerably from year to year in the United States on account of the
causative agents (that is, humans versus lightning), although
burned area caused by lightning fires is considerably more variable than that of human-caused fires. We would expect the same
variability to characterize fire activities in regions both inside and
outside the United States, with episodic peaks and lulls in fire activity, and with some years near average in ignitions and area
burned. Although global estimates of annual burning will vary,
1,300 billion acres (526 billion ha) is probably a reasonable estimate for the extent of worldwide burning in wildland areas,
mostly all the result of human causes.

As might be expected, area burned is not a perfect indicator
for all the costs, damage to natural resources, and ecological effects of forest fires. Even so, area burned is easily measured and
sometimes becomes the best proxy for the economic and ecological impacts of forest fires that are difficult to estimate. In the
United States, federal agencies of late have started to post wildfire suppression costs (www.nifc.gov/stats), especially as public
concerns have grown about the high expense of firefighting.

Generally larger fires are more expensive than small fires, although the costs per unit (per acre or hectare burned) are higher
on small fires. Even so, suppression expenditures alone will be a

poor reflection of all costs associated with wildfires and their management during any year. Figure 5–4 shows U.S. federal wildfire suppression costs over time, in this case for the period 1994–2000. Clearly, the USDA Forest Service expends more than other federal agencies combined from year to year, followed by the Bureau of Land Management (BLM). These agencies protect more public land than the other federal agencies, so we would expect higher total expenditures. In 1994, combined federal suppression expenditures for the first time exceeded $1 billion. By the year 2000, a single agency (USDA Forest Service) exceeded the $1 billion expenditure benchmark. Since then, annual federal agency expenditures routinely exceed $1 billion. However, those estimates may be misleading, since suppression costs by and large reflect expenditures on large fires—spending prior to a fire for staffing, equipment, or other preparedness measures may not be included. What's more, spending by state, local, and private entities will not

Figure 5-4
Suppression Expenditures by Federal Agency, 1994–2000

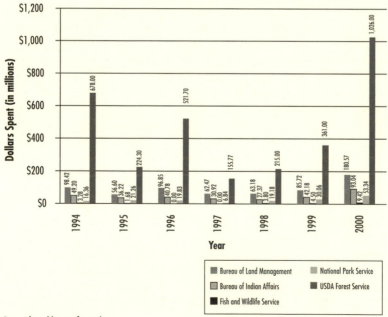

Source: http://www.nifc.gov/stats

Note: Figures are unadjusted by inflation.

be reflected. Moreover, natural resource (that is, timber, recreation, water, wildlife, and range) damage, ecosystem impacts, and business losses are rarely included in cost-accounting schemes.

Variation in expenditures can be explained by differences in agency missions, land base, fire loads, and recording standards, as well as numerous other considerations. Table 5–1 provides a sense for the differences in agency mission and land base, but differential fire loads are difficult to assess because a large fire may involve multiple jurisdictions and cross numerous property boundaries as it spreads. Typically, expenditures will be assigned to the responsible agency at the location of the fire's origin, but eventual costs may be shared with other organizations, depending upon how the fire spreads and cooperative arrangements among participating entities. Sometimes cost assignments and re-imbursements among agencies will require months and even years to resolve, depending on fire complexity and spread patterns. For example, if state and private lands are involved, the process could drag on even longer. Thus cost estimates probably need to be viewed with considerable caution. In fairness, the federal agencies protect vastly different pieces of real estate, and costs would be expected to vary among timbered national forests of the USDA Forest Service; the shrublands of the BLM; scenic

Table 5-1
Agencies, Land Base, and Primary Management Units

Agency	Land area, in million ac (million ha)	Primary management units
USDA Forest Service	191 (77.3)	155 national forests and 20 national grasslands
USDI Bureau of Land Management	270 (109.3)	Multiple use areas within 50 states
USDI National Park Service	80 (32.4)	166 national parks, monuments, preserves, and seashores
USDI Fish and Wildlife Service	95 (38.5)	540 National wildlife refuges and thousands of small wetlands and special management areas.
USDI Bureau of Indian Affairs	56 (22.7)	Tribal and Indian Trust lands

Note: Differences in land base and primary management units help explain some, but not all, variability in annual expenditures by U.S. federal agencies.

Source: http://www.nifc.gov/stats

vistas in national parks and monuments; wetlands in wildlife refuges; and native resources on Indian trust lands. In addition to the differences in values protected, fire behavior and effects will vary widely among the various fuel types found within each agency's jurisdiction. Differences in expenditures may also reflect agency missions and management practices—and the extent to which agencies have come to grips with managing fire problems in the past.

Information on prescribed fire acreage and costs is hard to come by, inasmuch as agency programs are relatively new and good record keeping hasn't always been standard practice. In the United States, federal agencies have been under increasing pressure to increase the use of prescribed fire and other fuel treatments (such as mechanical thinning), especially after the bad fire years in 2000 and 2002. Available information for the years 1995–2000 (www.nifc.gov/stats) reveals that federal agency use of prescribed fire averaged about 1.4 million acres (567,000 ha). Prescribed fire costs ranged from about $17 per acre ($42/ha) for the U.S. Fish and Wildlife Service in 1998 to $81 per acre ($201/ha) for the National Park Service in the same period, with expenditures for the other federal agencies falling between those bounds (Omi 2004).

Compared with the agency land bases in Table 5–1, the annual prescribed fire accomplishment is fairly low, considering the total area under each agency's jurisdiction. In fact, the annual area burned by both wild and prescribed fire on federal lands constitutes about 1 percent or less of total agency land bases. The difference in prescribed fire costs between agencies reflects a variety of factors, including type and size of land base, agency mission, and workforce expertise and experience. For example, much of the Fish and Wildlife Service land base includes wildlife refuges and wetlands in the Southeast, where fire use is ingrained in the local culture. By contrast, the National Park Service's prescribed burning efforts are concentrated in high-visibility parks such as Yosemite, Grand Canyon, and Sequoia–Kings Canyon; costs are higher in such locales in part because of the high values at risk and limited opportunities for burning as a result of fire danger and air quality concerns. Furthermore, cost estimates for prescribed fires reflect expenditures for execution but not necessarily planning, preparation, and data collection—or the costs associated with escapes, which would show up as wildfire expenditures. Rideout and Omi (1995) have described how numerous reasons can account for the variability in unit cost estimates for prescribed burning, including the size of the burn project, objectives, the complexity of the project, and fuel types. Generally speaking, the larger the prescribed burn project area, the lower the unit costs.

Fire suppression costs can range anywhere from five to a hundred times those of prescribed fires. Prescribed fire and fire suppression are not substitutes for each other, however, so the comparison is not totally meaningful. On the other hand, the pattern of recent wildfires throughout the arid West has left an impression that wildfires are not manageable, especially under extreme burning conditions. And in fact, wildfires have reduced fuel loads in more areas in recent years than will likely ever be achieved by intentional fuel hazard reduction. Within that context, then, prescribed fires not only cost considerably less than fire suppression but, overall, also cause less damage to organisms and ecosystems; in fact, depending on burning conditions and objectives, prescribed fires may be beneficial and perhaps required for restoring the health of some ecosystems. Since prescribed fires can be used to reduce fuel hazards and the severity of eventual wildfires, society has a fairly clear-cut choice to make regarding fire management on public lands—that is, application of fuel treatments such as prescribed fire at relatively low costs, versus massive expenditures and damaging wildfires, especially as witnessed in bad fire years such as 2000 and 2002. Those two years were widely acknowledged as two of the worst to date in terms of costs and losses resulting from U.S. wildfires.

Although most fires remain small and result in relatively insignificant impacts, occasionally fires grow large and cause significant damage, including loss of life (as discussed in Chapters 1 and 2). In some years, climate and fuel conditions will contribute to numerous large fires regionally. Thus a plot of fire frequency against size would have the shape conveyed in Figure 5–5, showing a high fraction of fires in the small size classes, decreasing proportions of midsized fires, and an even smaller ratio of large fires (to the total number of fires). The relatively few large fires (say in excess of 1,000 acres) cause the bulk of natural resource damage and housing losses.

Thus relatively small fires occur frequently, while larger fires are relatively infrequent. For example, small fires (say, less than 10 acres, or 4 ha) might result from 65 to 75 percent of all ignitions, while extremely large fires might constitute 2 to 3 percent

Figure 5-5
Hypothetical Fire Size versus Fire Frequency
Hypothetical fire size versus fire frequency distribution, showing
the high relative portion of small fires and low portion of
very large fires.

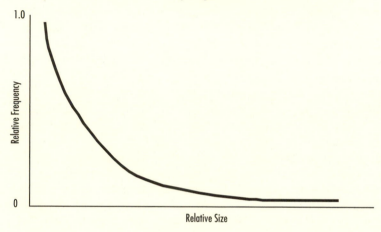

of the total—depending on location and climate. The right-hand tail of the curve in Figure 5–5 represents the low relative frequency of large megafires—those that typically receive disproportionate coverage in the news media during a fire season.

Some of the most notable large fires have been discussed previously in Chapter 3, in which we noted the better known fatality fires. Even so, fatalities do not occur only in large fires, or only in heavily timbered forests. In fact, Wilson (1977) found that most fire fatalities have occurred in flashy fuels or on the periphery of larger fires. Fires in light fuels can be deceptively dangerous, since spread rates are high and fires in such fuel types can change direction and accelerate quickly as the wind shifts. Fire fatalities represent yet another cost (perhaps the highest) to society. Figure 5–6 plots firefighter fatalities by decade for the period 1900–2002 in the United States. Firefighter and public safety has become a top priority for federal land managers, yet fatalities persist at about 150 per decade.

Although fatalities occur in some large fires, smaller fires burning in lighter, flashy fuels also can threaten firefighters. Most fatalities occur when firefighters are surprised or caught unawares by sudden, shifting winds or increases in fire intensity (sometimes called "blowup" conditions). Fire blowups can trap

Figure 5-6
Firefighter Fatalities, 1900–Present
The absence of fatalities during 1911–1920 may be due to poor
record-keeping (common during the early periods of
US fire management)

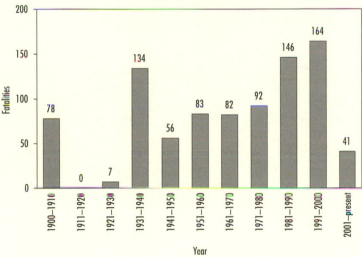

Source: http://www.nifc.gov/stats

unsuspecting or weary firefighters, including seasoned veterans
with many years of experience. For example, the fourteen fatalities on the 1994 South Canyon fire in Colorado included members
of elite smokejumper and hot-shot crews who were overrun by a
sudden and unexpected surge in fire intensity. Surprisingly anal-
ogous circumstances contributed to the deaths of thirteen Montana smokejumpers in the 1949 Mann Gulch fire. More commonly, firefighting crews with less training and experience may
experience fire entrapments or suffer fatalities. Inadequate crew
cohesion, poor communications, and lack of awareness of hazardous situations are among the contributory causes (Alexander
2004). Thus fatalities and injuries can strike all firefighters, from
the most experienced to the least well trained. No amount of training or supervisory reminders about safety seems to matter when
firefighters become caught in a sudden blowup—sadly, the best
tactic is to avoid being in the wrong place at an inopportune time.

As noted in Chapters 3 and 4, the chronology of costs and
losses (including fatalities) from wildfires has influenced the evolution of forestry and natural resource policy. For example, the

policy review that followed the 1994 South Canyon fire established priorities for firefighting (in decreasing order): firefighter and public safety, private property, then natural resources. Typically, bad fire seasons and fatality fires spawn numerous reviews and extensive soul searching, and they may result in new legislation, national initiatives, and new local laws and ordinances, mostly in reaction to casualties and other losses. Concurrently, agencies charged with managing fires on public and private lands have developed sophisticated technologies and information systems to contend with perceived threats to society from wildland fire. Not surprisingly, large forest fires have spawned a substantial private industry that includes aircraft (both fixed and rotary wing), fire suppression supplies and equipment, and an enormous amount of support services (for example, kitchen and sanitation services, radio communications, and monitoring equipment). At the same time, a growing body of scientific evidence indicates that all-out suppression of wildfires may not be desirable or sustainable in some ecosystems (see Chapters 1 and 2).

Several caveats are necessary in the interpretation of historical fire records. Fire records have been collected by public agencies for only a relatively short time—say, fifty years or less. By contrast, fires have been burning for thousands of years, so the available records provide only a brief statistical snapshot, the clarity of which becomes fuzzier and more incomplete as you venture further back in time. Recording standards change over time, as well as geographical coverage and archive technologies. In some cases, the length of record may be extended considerably by reliance on other historical sources, such as datable fire scars on trees or charcoal sediments in lakes, which may go back several centuries or more. Although extremely useful for resurrecting fire histories, including estimates for fire recurrence intervals, these other techniques possess their own methodological difficulties and can be quite time consuming. We will discuss these techniques in more detail later, but first we will examine some of the fundamental outcomes from fire in the forest.

Forest Combustion

Forests, shrublands, and grasslands have burned for millennia as a result of natural causes and human firing practices. Aborigines must have recognized fairly early that fire could be both a useful tool as well as a destructive environmental change agent. As civilizations developed, the recognition likewise must have developed that some ignitions provided useful benefits, such as creating clearings for agriculture, improving wildlife habitat, or reducing fuel hazards, whereas other fires could not be allowed to burn unimpeded. Over time the benefits of agricultural fire were transformed for use in industrialized societies.

The mechanics of suppressing fire are well known to firefighters. The classic fire triangle consists of fuel, oxygen, and heat (Figure 5–7). In the forest, fuel refers to live and dead vegetation that will burn. As we are taught in grade school, we can extinguish fires by breaking or removing one of the legs of the fire triangle, removing the possibility of combustion. Similarly, firefighters are taught that water and soil can be thrown at a fire to remove heat and oxygen, while building a fireline will create a barrier that essentially removes fuel ahead of the fire. What's more, a fire can be prevented if one of the legs of the triangle is not permitted to mix with the other two. For example, oxygen and fuel in the absence of heat is not a combustible situation. Moreover, variations in fire behavior result from the changing mix of fuel, oxygen, and heat within the combustion zone.

Figure 5-7
Classic Fire Triangle
The classic fire triangle consists of fuel, oxygen, and heat

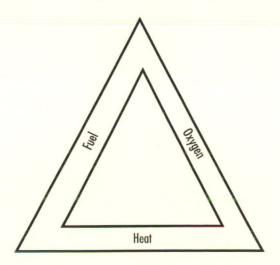

The physical and chemical properties of wildland fuels contribute to their combustibility. Physical properties include attributes of the fuel particles, such as their size and density, as well as the horizontal and vertical arrangement of particles throughout the fuelbed. Fuelbeds that are too tightly or too loosely packed with fuel particles may not burn efficiently, just as high particle moisture content will dampen combustion. Terpenes, oils, and other volatile chemicals (so-called ether extractives), evident in pine or chaparral fuels by their distinctive aromas, will facilitate active burning.

All vegetative biomass that will burn in the forest consists of complicated arrangements of cellulose, hemicelluloses, and lignin, incorporated within a matrix of extractives and mineral constituents (Chandler et al. 1983a). One ton of wood consists of 50 to 75 percent cellulose and hemicellulose, with a heat content of 10,500 Btu/lb (24.40 MJ/kg); 15 to 35 percent lignin (heat content 10,500 Btu/lb or 24.40 MJ/kg); 0.3 to 15 percent ether extractives (terpenes and resins, heat content 13,900 Btu/lb or 32.30 MJ/kg); and 2 to 4 percent ash (inorganic compounds that do not burn) (Deeming 1988). The combustion process can be represented simplistically by equation (1), which shows the oxidation of a glucose molecule (a basic chemical unit of cellulose, the primary constituent of wood).

$$C_6H_{12}O_6 + 6O_2 \rightarrow 6CO_2 + 6H_2O + heat \qquad (1)$$

Woody fuels and plant material in the forest are more complex chemically than glucose, but equation (1) represents conceptually the combustion process that we observe in the field. Thus oxygen combines with woody fuels to produce carbon dioxide (CO_2), water (H_2O), and heat. The amount of heat given off in the combustion reaction is of interest because many fire effects will be related to the heat output from a fire. The heat of combustion by various materials can be determined experimentally using a bomb calorimeter in a laboratory. For example, Table 5–2 shows a sampling from previous experiments, demonstrating little variation between types of wood with higher energy contents for pine pitch and eucalyptus oil, as might be expected because of their higher volatility.

Actually, combustion in the wild is not an especially efficient process, so the proportions of carbon dioxide, water, and heat

Table 5-2
Heat of Combustion from a Variety of North American and Australian Fuels

Fuel	Heat of combustion (MJ/kg)
Pine wood	21.28
Oak wood	19.33
Poplar wood	18.22
Eucalyptus (average of 9 species)	19.98
Pine sawdust	21.74
Spruce sawdust	19.65
Pine pitch	35.13
Eucalyptus oil	37.20

Source: After Whelan 1995.

produced will vary considerably, as will the flame and gaseous temperatures emitted from burning fuels. For example, the actual heat yield per kg of various burning fuels will be considerably less than that shown in Table 5–2, largely a consequence of the heat required to drive off and vaporize moisture in the fuel particles and resultant incomplete combustion.

Furthermore, numerous other by-products may result from incomplete combustion and chemical reactions with constituents of burning fuel or in the air (for example, nitrogen in plant parts, in soil litter, or in the atmosphere will react with oxygen and result in nitrogen oxides). Other by-products include carbon monoxide (CO), particulates (a major air pollutant), and gaseous hydrocarbons, some of which are toxic and carcinogenic. CO in particular is a by-product of incomplete combustion and can be lethal, except that its concentrations dissipate rather quickly with increased distance from flaming and smoldering fuel sources. Approximate by-product proportions are portrayed in Figure 5–8, although proportions would be expected to vary depending on fire intensity and the completeness of combustion. Also, total suspended particulates are probably under-represented in the figure, and concentrations could be several orders of magnitude larger. Woody fuels also produce sulfur oxides in small amounts, a major constituent of urban-industrial pollution. Trace amounts of oxidants that react with sunlight and are found in photochemical smog have been found in wood smoke. The longevity of photo-reactive substances found in smoke is poorly understood but probably short-lived. Large amounts of ozone have been noted in

Figure 5-8
Emission or By-Product Proportions
Fractions may vary depending on combustion efficiency and fire
intensity. Total suspended particulate fractions are probably under-
represented. Also, trace proportions of sulfur oxides and other
oxidants are not shown.

Emission proportions (approximate)

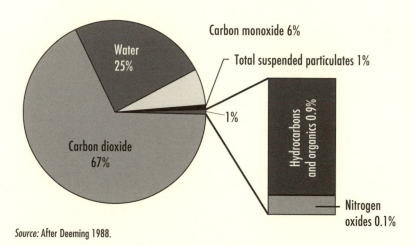

Carbon monoxide 6%

Total suspended particulates 1%

Water 25%

Carbon dioxide 67%

1%

Hydrocarbons and organics 0.9%

Nitrogen oxides 0.1%

Source: After Deeming 1988.

smoke plumes aloft, but little data link fire emissions with increased ground-level ozone concentrations (Sandberg et al. 2002).

At present, particulates are the major pollutant of concern
from wildland fires in the United States (ibid.), although globally
greenhouse gas emissions (for example, CO_2) are also of concern.
Particulates include soots, tars, and condensed organic substances entrained in the smoke plume, which may range in size
from 0.01 to 15µ (µ = 1 micron, or 10^{-6} meter, roughly the width
of a fine human hair). The approximate proportions of particulates of different size classes in smoke from wildland combustion
are shown in Figure 5–9. The larger particulates scatter light and
can reduce visibility. Smaller particulates may be suspended in
the air that we view and breathe. Two classes of fine particulate
matter have generated great concern from the standpoint of air
quality and human health: PM10 (particulate matter less than 10µ
diameter) and PM2.5 (particulate matter less than 2.5µ diameter).

Figure 5-9
Most particulates entrained in smoke are in the finer size classes.

Particulate proportions

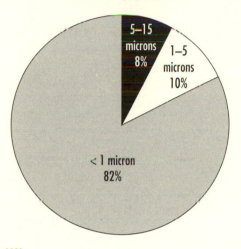

Source: After Deeming 1988.

Consistent with Figure 5–9, later studies (for example, Ward and Hardy 1991) indicate that 90 percent of all smoke particles emitted during burning are PM10, and 90 percent of PM10 are PM2.5 (ibid.). Thus the largest proportions of particulate diameters in smoke are found in the smallest size classes. Prolonged exposure to such by-products can affect human health, especially for people with chronic pulmonary disorders.

Combustion Stages

The oxidation of woody biomass occurs all the time in the forest, as fuels are naturally decomposed by microbial activity. Combustion hastens natural decomposition of organic biomass into its constituent parts, sometimes in spectacular displays of heat, light, and sound. Burning produces relatively large amounts of energy in two forms: convection (from the motion of air) and radiation (infrared and visible light). Relatively smaller proportions of energy are transferred by conduction (heat transmission through a solid material). The combustion process starts as an endothermic reaction with heat provided by a spark or other ignition source, then becomes a self-sustaining exothermic reaction (1) until extinction occurs, when fuels become unavailable because of changes in environmental conditions or human suppression. In between ignition and extinction, the fire will go through several stages, including preheating of unburned fuels, followed by flaming, glowing, and smoldering combustion (Table 5–3). During preheating, heat is required to drive off moisture and convert solid fuel to volatile gases, otherwise known as pyrolysis. During flaming, the combustion reaction becomes exothermic, with the burning of volatiles and flame temperatures in excess of 500–600° C (Agee 1993). Glowing and smoldering involves the burning of solid carbon. Extinction occurs after fuels are cooled or no longer available for combustion. Particulate production is approximately 2.5 times higher during smoldering than during flaming, because of the greater incompleteness of combustion during latter stages of the fire (Deeming 1988).

Table 5-3
Fire Stages, Burning Characteristics, and Approximate Temperature Ranges

Combustion stage	Description	Temperature range
Preheating	Thermal dehydration and degradation of solid fuels ahead of the fire. Lignin and cellulose gasify at 150° C to produce combustible gases.	< 300° C
Flaming	Gaseous volatiles ignite when sufficient oxygen is present and carry flame, preheating unburned fuels by radiation, convection, conduction, and direct flame contact.	300–600° C, with spikes to 1200° C in crown fires
Glowing and smoldering	Burning of solid charcoal (carbon), sporadic flaming due to insufficient oxygen to support flaming or when pyrolysis occurs too slowly to supply sufficient volatiles because of lower temperatures.	~500° C
Cooling and extinction	Decrease in heat emitted until residual fuels are at ambient temperature; ash layer persists where complete combustion has occurred.	20–500° C

Source: After Agee 1993 and Deeming 1988.

These stages exist at different points along the fire's perimeter as the fire spreads through the forest and across a landscape. The fire may develop a cohesive flaming head, emitting intense temperatures (brief peak temperatures in excess of 1,000–1,200°C in raging crown fires), bright and glowing light, and loud sonic roars—or else grow in sporadic fits and spurts, in response to changing fuel availability (as affected primarily by changing winds and fuel moisture contents). Some of the most dramatic fire effects may occur after passage of the flaming front, as glowing and then smoldering combustion may last for hours or even days, killing stems and roots and consuming ground cover. Eventually, these portions of the fire become quiescent or cold as the fire dies out or is extinguished. In general, the most active burning conditions occur at the fire's head (or front), with lesser activity along the flanks and to the rear. If intensities are great enough at the active flaming front, shrub or tree crowns may become engaged in the combustion process.

At all stages prior to cooling and extinction, vulnerable houses and structures exposed to heat may be at risk. Risks to houses may seem greatest during flaming stages, especially when burning embers may shower down on roofs or other combustible surfaces ahead of a fire's flaming front. Even so, some structures may be ignited by creeping surface fires if enough heat is generated. Houses may burst into flames after decks or other combustible attachments are subjected to prolonged heating, or if a fire reburns an area because of shifting winds. In these respects, houses and adjoining structures are just like other fuels in the forest.

We also can think of forest combustion as the rapid breakdown of plant biomass, accompanied by the release of the chemical energy that the plant had originally converted from solar radiation via photosynthesis (Agee 1993). Trollope (1984) compared those two reverse processes as follows, using a simplified representation for a carbohydrate to represent the cellulose molecule $(C_6H_{10}O_5)_n$ as a result of photosynthesis (equation 2) and an oxidant in combustion (equation 3):

$$CO_2 + H_2O + \text{solar energy} \rightarrow (C_6H_{10}O_5)_n + O_2 \tag{2}$$

$$(C_6H_{10}O_5)_n + O_2 + \text{heat} \rightarrow CO_2 + H_2O + \text{heat} \tag{3}$$

During a fire, the chemical energy stored in forest fuels is converted to radiant, kinetic, and thermal energy (Byram 1959).

As a fire grows it develops an anatomy that reflects differential rates of energy conversion along its perimeter. Fire effects will vary at different segments of a fire's perimeter, as the temperatures emitted (and duration of heat exposure) can change dramatically, depending on the fire's environment and anatomy. Temperatures will likely be highest at the flaming front, while durations may be longer in the back or flank of the fire. Eventually even the hottest portions of a fire will cool. Fire effects will also differ with anatomy, since the heat generated and resistance of affected plants and organisms will produce a mosaic of patchy burn patterns.

Similarly, mechanisms of heat transfer will vary depending on fire spread in relation to terrain slope and wind. Thus a fire burning on flat terrain with no wind will spread primarily by radiation, while a fire burning with the wind or up a slope (head fire) will spread by radiation, convection, and flame contact. By contrast, a fire spreading into the wind or down a slope (backing fire) spreads primarily by radiation (Rothermel 1972).

The amount and extent of flaming will depend on the rate of decomposition from solid fuel into volatile gases, previously described as pyrolysis. These gases burn during flaming when mixed with oxygen. The fire's intensity and flaming will depend on the rate at which solid fuel is preheated and pyrolyzed, then mixed with surrounding oxygen. Thus the rate of pyrolysis is highest when large amounts of fuel are burning rapidly and is lowest during the smoldering combustion preceding extinction.

Pulses in fire growth can be caused by environmental changes (that is, higher or gusting wind speeds, differential solar heating or shading of fuels) or by encounters with fuel jackpots (for example, piles of flammable branches and twigs, or needles draped over understory bushes). Fires usually slow down after sundown on account of the cooler temperatures and higher relative humidities—but they may roar back to life the next day, as the nighttime inversion (warm air over cold) dissipates in response to solar heating. Passage of thunder clouds or convection columns from a fire can create dangerous updrafts and downdrafts that push and pull at a spreading fire, causing sudden changes in the directional rate of growth.

Chemical properties that contribute to flammability include cellulose/lignin proportions, ether extractives; mineral constituents (except silica), and moisture contents generally inhibit combustion. Technically, the amount of oxygen available for combustion depends on the elevation and air pressure, though oxygen availability rarely limits forest combustion. More important, the speed and direction of winds (including those that are induced by the fire itself) provide ample oxygen circulation within fuel arrays to support the combustion process. Higher air and fuel temperatures facilitate more active burning, while higher humidities attendant with cooler temperatures will dampen combustibility of finer fuels.

Firefighters use knowledge of fire growth processes to suppress an ongoing fire by building firelines around a fire's perimeter to isolate fuels from heat. Spraying water or fire retardant on the fire can reduce heat. Total immersion of the fire in water or soil will rob it of oxygen, though total O_2 deprivation is not practicable for suppressing a large fire. Instead, firefighters are taught to contain a fire by building a line completely around a fire's perimeter, starting from an anchor point and eventually encircling the fire, incorporating natural barriers where available, such as lakes, rivers, or boulder fields. A fire surrounded by fireline is considered contained. While building fireline, firefighters may engage in hotspotting, by throwing soil at the base of flames to remove heat and oxygen; firefighters also may call in air support to drop water or fire retardant on active fire edges. During the mop-up stages of a fire, firefighters will remove oxygen and heat from burning areas by mixing soil and water with embers or pockets of burning fuels.

Fire Environment

Fire managers recognize the importance of the fire triangle (Figure 5–7) in explaining the ignition (and control) of forest fires. Firefighters also are trained to consider the various alternatives for suppressing heat transfer to unburned fuels that lie in the path of an oncoming wildland fire, using principles embodied in the fire environment concept. The fire environment is the complex of fuel, topographic, and air mass constituents that influence the inception, growth, and behavior of a fire (Countryman 1972). The fire environment dictates fire behavior through dynamic changes over space and time in fuel, air mass, and topography. The elements of the fire environment interact with one another, often in chain reaction sequences that can dramatically alter the course of a fire. For example, air mass changes (such as wind speed and direction) can induce dramatic changes in fire behavior, although those effects may be most pronounced in mountainous terrain where broken topography and fuel type changes are likely.

Table 5–4 depicts important fuel, weather, and topographic influences on fire behavior. Dramatic changes in fire behavior can result from changes in any one of the elements in the fire environment. Furthermore, most of these will vary simultaneously as a fire moves through an area. Moreover, they may interact, increasing or restricting fuel availability and the problems encountered by firefighters. For example, a high wind speed may enable fires to spread in moist fuels. Or, in the Northern Hemisphere similar fuels will burn more dramatically on south than on north aspects, because of solar heating. Generally, self-sustaining combustion occurring in the forest will create problems for firefighters if the relative humidity falls below 25 percent, if the moisture content of fine fuels (less than 1 inch, or 2.54 cm, in diameter) averages around 5 percent, and wind speeds exceed 10 mph (16.7 km/hr). Extreme fire behavior may be observed if moisture levels fall below (or winds exceed) these thresholds, although other contributors to extreme fire behavior include continuous fuels linking the surface and crown fuel strata (that is, ladder fuels), or fuel jackpots (piles of flammable material).

Fire managers focus particular attention on fuels because weather and topography are less easily manipulated. Of the elements of the fire environment triangle, only fuel can be manipulated in advance to render future ignitions more manageable. However, on an ongoing wildfire, experienced firefighters will make mental notes of slope, aspect, and time of day because of

Table 5-4
Indicators of a Fire's Environment

Fuels	Air mass	Topography
Fuel moisture	Wind speed and direction	Slope steepness
Vertical arrangement	Relative humidity	Slope position
Fuel loading	Precipitation	Aspect
Compactness	Temperature	Elevation
Size (i.e., timelag) and shape	Atmospheric stability	Shape of country
Horizontal continuity		Canyon shape
Chemistry		Canyon width
		Ridges
		Flat terrain

A fire's environment consists of fuels, air mass, and topography. Indicators may interact with one another (e.g., wind and fuel moisture), thereby influencing the availability of fuels for combustion and creating different challenges for firefighters.

their potential impacts on fire behavior and growth. Changes in fire environment during a day can be just as important to a fire-fighter as the more obvious signals, such as the jetlike roar of an approaching crown fire. Subtle shifts in wind speed or direction can portend imminent changes in burning conditions. In addition, the fire environment will dictate the ecological consequences of fires by controlling the rate of spread, intensity, duration, and severity of burning.

Few people realize that a fire seldom consumes all burnable material or scorches every square meter of ground surface. Even the most extreme fires will create a patchy mosaic of burned and unburned areas, because of the variability of fire spread and flaming, fuel discontinuities, weather changes, and natural barriers. Combustion availability is determined by fuel, weather, and topographic factors included in Table 5–4. Thus fire scientists and managers distinguish between *fuel amount or quantity* and *fuel available for combustion* when discussing the fire potential for an area.

Fuel quantity is expressed in metric tons/ha or tons/acre, or sometimes kg/m^2 or lb/ft^2, or units of mass per unit area. Fuel availability refers to the portion of total fuel that actually burns in a fire, and will depend on the above fuel descriptors (Table 5–4) as well as weather and topographic gradients. Fuel availability varies by plant community type, fire environment, and disturbance history for an area. Under low flammability conditions, only surface fuels and low vegetation may actually combust. Moderate burning conditions may ignite downed logs and standing dead snags and scorch the lower tree crowns. Under extreme burning conditions, entire tree crowns may be enveloped in flames. All the fuels within a grassland may be available for combustion, whereas only a small portion of an old-growth forest may actually catch fire, except under extreme drought conditions. Fuels vary in their availability for combustion in response to changes in the fire environment over space and time. Fuelbeds and their constituent fuel particles will change throughout the day and seasonally in terms of the ease with which they ignite and continue to burn. Thus high winds and low relative humidities will augment fuel availability, just as cold rain or snowstorms may knock the life out of a fire by rendering all fuels unavailable. Prolonged drought increases fuel availability, especially among the larger diameter fuels whose moisture levels may stay uniformly low even as the finer fuels are responding to diurnal humidity fluctuations. During the summer, fuel moisture may change most quickly, especially in fine fuels such as dead grass; the moisture content of larger fuel particles (such as logs) may change rather slowly with time. Consequently, fuel moisture is a major determinant of variations in fire behavior and potential fire hazard.

Dead logs, branches, and twigs are hygroscopic, in that they exchange moisture with the environment. The rate of moisture change in a fuel depends on the relative humidity (amount of moisture in the atmosphere compared with saturation conditions) of the air immediately surrounding the particle and the particle size. The common measure of fuel particle size is its surface area to volume ratio, which can be likened to a comparative measure of the fuel "skin" (surface area) and amount of space within the skin (volume). Small diameter and thin fuels have a relatively larger surface area to volume ratio and can respond rapidly to changes in environmental moisture. By contrast, large logs have a small surface area compared with volume, and internal moisture changes occur more slowly. A large log with high moisture content, for example, may require months before its interior becomes dry enough to burn.

If a desiccated piece of fuel (say a dead blade of grass) is placed in a fixed, moist environment, its moisture will increase rapidly, then slow, then cease changing. The moisture in the fuel is then at equilibrium with the environment—or at its equilibrium moisture content (emc). A reverse trend would be observed if a moist fuel were placed in a dry environment—that is, a rapid decrease in moisture content, then a slowing in the rate of decrease, and eventual equilibrium (Figure 5–10). The rate of increase (or decrease) and corresponding equilibrium point will depend on ambient temperature and relative humidity.

The rate of moisture change in a fuel particle in response to increases or decreases in relative humidity is called its timelag, a convention adopted for fire danger rating and fire behavior prediction purposes. Technically, one timelag is defined as the time required for the particle to reach $(1-1/e)$ percent, or roughly 63 percent, of the equilibrium moisture content, where e is the base of the Naiperian logarithms. Dead grass, foliage, and twigs less than ¼ inch (0.64 cm) diameter are called 1-hour timelag fuels, an indication of the approximate time required for a particle to approach equilibrium with its environment in response to a change in relative humidity. Similarly, branches between ¼ inch and 1 inch (0.64 cm and 2.54 cm) diameter are called 10-hour timelag fuels; they tend to follow diurnal trends in relative humidity and

Figure 5-10
Pattern of Fuel Moisture Content
Hypothetical changes in a fuel moisture content over time in a drying
environment, under constant temperature and atmospheric pressure.

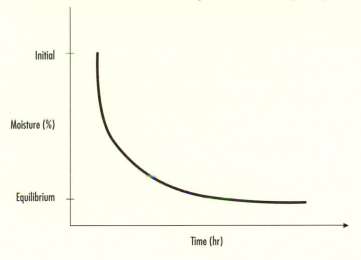

temperatures, rather than short-term fluctuations (Countryman n.d.). The 100-hour timelag fuels fall between 1 inch and 3 inches (2.54 cm and 7.62 cm); 1,000-hour timelag fuels exceed 3 inches (7.62 cm) in diameter.

The arrangement of fuel pieces within the fuelbed will also influence flammability. A highly compact fuelbed with a dense arrangement of fuel pieces may not allow sufficient oxygen circulation to promote active burning. Similarly, a sparse fuelbed may not have enough flammable biomass to support combustion. Horizontal continuity is required for flames to preheat unburned fuels ahead of a fire, just as fuels arrayed vertically will allow heat and flames to reach into a forest canopy and burn as a crown fire. Foliage and needles that contain oils and terpenes may enhance combustion, especially on a hot summer day when such fuels will emit gaseous volatiles because of solar heating.

The quantity of fuel (or loading), in tons/acre, metric tonnes/ha, or kg/sq m, will be important to describing fuelbed flammability. Fuel quantities can vary from very little to all the standing trees and canopy foliage in a forest. Whether an ignition spreads through the forest will depend on the size of fuel particles, moisture content, and spatial arrangement. Although vegetation biomass increases predictably with time as a result of ongoing photosynthesis, temporal fuel biomass changes can be irregular because of tradeoffs between annual fuel increment versus decay and environmental properties that affect fuel availability (Brown 2000).

Atmospheric stability, which measures the propensity of air to move vertically in response to a lifting (or lowering) force, indirectly influences fuel availability. For example, an unstable atmospheric environment (usually expected on a clear, hot summer day) will allow greater convective lift, raising flame heights and facilitating formation of a well-formed smoke plume or convection column above a fire—thereby supporting extreme burning conditions as ladder and crown fuels become available for combustion. By contrast, a stable environment (such as an early morning inversion created when a warm air layer overlies cold air) will constrain burning by resisting the convective lift from a fire.

In the mountains topography can be rugged and broken, with abrupt changes in aspect and slope angle. A fire reacts to steeper slope with increases in spread rate, flame length, and overall heat. As slopes increase, flames are bent closer to unburned fuel ahead of a fire, which promotes preheating of fuels and faster spread rates. As a very rough guideline, managers expect that a slope increase by 20 percent may approximately double the fire's rate of spread, all other factors held constant. The actual increase in spread rate may depend on the fuel type and whether the flames attach to the slope. Andrews and Chase (1989) note that flame attachment causes convective gases and smoke to flow up the slope, close to the surface, rather than rising vertically as on shallower slopes. When flames rise vertically, spread rates may be lower and overstory trees may experience higher scorch heights, except upslope where the fire encounters a ridgeline.

Solar heating, temperature/relative humidity regimes, and resultant fire behavior also will be influenced by slope position, aspect, and elevation. For example, a fire burning at the base of a south-facing slope in the early afternoon can create problems for firefighters, especially if high air temperature and low relative humidity combine with increasing upslope winds to push a fire uphill. Terrain shape interacts with environmental conditions (such as gradient winds and solar heating) to promote or retard fire behavior. Steeper, narrow canyons will promote dramatically different fire behaviors than will wider, flatter canyons or valleys.

As another example of fire environment interactions, consider the effects of topographic aspect on burning conditions in a hypothetical area in the Northern Hemisphere (Figure 5–11) during early summer. Fuel temperatures depend partly on air temperature and aspect. A slope facing east warms first and reaches its maximum early in the day. A south-facing slope reaches its maximum around four hours later, and that maximum is higher than on the east aspect. A west-facing slope reaches its maximum temperature even later, exceeding those of the east and south slopes. The temperature maximum on the north-facing slope (early PM) tends to be lower than the other aspects during the day.

As a slope is heated, air and fuel temperatures increase. In turn, relative humidity decreases, affecting the moisture of

Figure 5-11
Hypothetical Surface Fuel Temperature Profiles
Hypothetical surface fuel temperature profiles on a clear day in the northern hemisphere (42° latitude), early July, slope 45 degrees.

Time-temperature Profiles

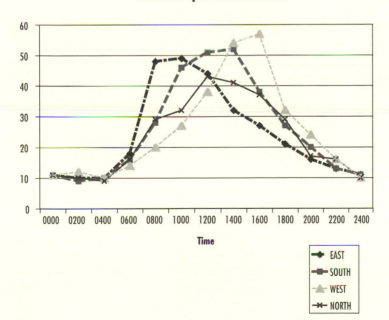

Source: adapted from Countryman 1966.

smaller-size fuel particles. Differential temperatures on adjacent slopes create pressure gradients, thus affecting local winds as air flows from areas of relatively high pressure to low. These winds will interact with the general winds in the area and be further modified by topographic shape and slope position. Cloud cover, slope percent, time of year, and latitude would change the pattern of the idealized temperature profiles in Figure 5–11. But suffice to say that fire initiation, growth, and extinction because of fire environmental influences would change considerably depending on aspect, time of day, and slope of the time-temperature curve in Figure 5–11—that is, whether temperatures had peaked or were yet to peak in the hours ahead. For example, a fire starting on a south-facing slope at 10 A.M. will confront favorable conditions for combustion during the remaining hours of daylight (the so-called burning period) than will a fire (also on a south-facing slope) ignited by lightning later that same evening—say, 10 P.M.—which will encounter a cooler, more humid environment. The morning fire will encounter upslope winds because of surface heating in the afternoon, which could lead to fairly dramatic burning conditions by late afternoon. By contrast, the evening fire would likely respond to cooling, downslope winds after sundown that might actually slow a fire's growth. Similar contrasts could be expected on west, east, and north aspects as well. These influences show up, sometimes dramatically, when a fire spreads from one topographic aspect to another, essentially coming under the influence of a different microclimate. For example, firefighters have been caught unawares when a benign surface fire burning on an east aspect around noon spreads to a south aspect and roars to life by mid to late afternoon.

Heating of the earth's surface contributes to local instability in the lower atmosphere, and the onrush of air to fill the vacuum created by convective lifting creates a surface wind that pushes a fire. As the lifted air is cooled it drops back to earth, only to be heated again at the earth's surface, perpetuating the convective cycle. Thus vertical motions (such as convective lifting) can be as important as horizontal surface winds in determining a fire's behavior. The stability of the atmosphere, and the resultant propensity of air to continue movement vertically, may either encourage or dampen upward air movement—and the growth and explosiveness of a fire.

This phenomenon is especially important for understanding variations in fire behavior in mountainous terrain. During the early morning, a fire may be discouraged from active burning by a surface inversion that persists in the mountains. Heat from the fire will generate vertical air movement, but the stability of the inversion will dampen the fire's spread and intensity. Later in the day, however, heating of the earth's surface by the sun will gradually increase atmospheric instability until the fire is no longer burning in a stable environment (around midmorning). At that point, fire spread can increase dramatically as the inversion is lifted or broken by convective currents arising from surface heating.

Thus heating (from the sun and from the fire itself) and air movements in the atmosphere are important drivers to a fire's behavior. Winds tend to be more turbulent and gusty when the atmosphere is unstable, creating more erratic and unpredictable fire behavior. Thunderstorms and well-formed convection columns with strong updrafts and downdrafts are symptomatic of unstable environments. Subsiding warm, dry air brought in from high-pressure areas to replace rising air in low-pressure areas can stimulate active burning conditions at night—sometimes making them the equal of daytime burning conditions.

Other air movements of importance are associated with general circulation patterns, including varying patterns of pressure systems and large air masses migrating across the earth's surface. Also, frontal winds (between two air masses) can be gusty and shift direction. Mountain topography can channel winds, alter wind patterns, and create eddies that alter fire behavior. Differential heating along coastlines and in mountain slopes and valleys can alter and interact with general winds. These are just a few examples of the different types of winds that can affect fire behavior and that fire managers must consider.

Overall, wind affects wildfire in numerous ways. Wind direction will mostly determine the direction of fire spread. Dry air movements can hasten the desiccation of forest fuels. Light winds can carry embers and firebrands to unburned areas. Winds increase oxygen flow into the fire and bend flames closer to unburned fuels ahead of the fire front. Winds also carry fire between shrub or tree crowns to sustain active burning. One of the most critical fire weather situations involves warm, dry winds (known as foehns, or gravity winds) that descend the lee slopes of mountain ranges in response to pressure gradients. These winds may be known locally as chinooks (east of the Rockies), Santa Anas (southern California), or east winds (Oregon), and they can be associated with extreme burning conditions. For example, the 2003 southern California fires were pushed downslope by Santa Ana winds (from the northeast), but additional problems were created when the foehn subsided and coastal breezes (from the west) pushed the fire back uphill in the San Bernardino Mountains. The crucial zone where coastal and foehn winds collide can present many problems for firefighters, depending on their relative strengths at surface elevations and aloft. Numerous firefighters have been killed in the turbulent wind zone where coalescing foehn and coastal winds can toss firebrands and create spot fires that entrap unsuspecting fire fighters.

Once ignited, a fire is continuously redefined and shaped by changes in its environment (for example, fuel, weather, topographic variations), so that the rate of spread, flame length, and growth indicators (such as perimeter increase and shape) can be expected to vary considerably. A fire reacts to changes in its environment by becoming more or less organized in its behavior. Generally, higher temperatures, lower moistures (in the air and in the fuels), and higher winds will result in higher rates of spread with visibly longer flame lengths. A well-formed convection column will be visible as heat from the fire is coupled with instability in the atmosphere (and a greater propensity for vertical motion). A quiescent fire burning with little organization on a south or west aspect at 10 A.M. may be transformed to a raging inferno by late afternoon, as heating from the sun aligns with upslope winds and terrain slope to contribute to a major fire run. Similarly, a benign flank may suddenly awake and become a wall of flame because of the passage of a cold front or sustained downdrafts from a thundercloud. Management challenges become even more complex when such major runs are spread over several drainages or on multiple adjacent fires.

Ultimately, the local environment will determine the frequency with which fire recurs in an area, as dead and live biomass respond to moisture and temperature gradients resulting in differential availability for combustion. The patterns of ignition over time may exhibit regularity because of lightning storm corridors or particular topographic landforms (such as ridges) that may be susceptible to fire starts. Human caused ignitions will often align along roadsides, railroads, or in high-use corridors around summer holiday dates. By the same token, fire effects can be expected to vary considerably from place to place over time, as will human perceptions of fire impacts and damage.

Fire Weather Observations

Short-term weather and long-term climatic changes profoundly affect a fire's behavior. Fatality fires often are associated with abrupt weather changes that surprise firefighters. Large wildfires occur during prolonged droughts. So weather observation and measurements are most important to the understanding and prediction of fire behavior. Drought (or water deficit), on the other hand, is more pernicious and sometimes difficult to identify until well after its onset. In this section we discuss the different measurements that can be used to estimate the short- and long-term influences of air mass on fire behavior. Short-term measurements included in standard fire weather observations include the following quantitative and subjective assessments of relevance to fire behavior forecasting on both wild and prescribed fires:

> Temperature: A measure of the degree of hotness or coldness of an object (not to be confused with heat, which is a form of energy). Dry and wet bulb temperatures are taken with a standard belt weather kit and are used to estimate dewpoint temperature, or the temperature at which the current environment becomes saturated.

> Relative Humidity: The ratio of the water vapor in a volume of air to the total amount of water vapor that the volume would hold at saturation (constant temperature and pressure). Relative humidity is computed from dry and wet bulb temperatures.

> Wind speed: A measure of the speed at which air is flowing past the observation point. Wind speed is commonly measured at different elevations above the earth's surface; fire behavior specialists pay particular attention to midflame and 20-foot (above the plant canopy) wind speeds, although winds aloft will also affect a fire's behavior. Estimates are derived from a variety of sources, including hand-held anemometers, nearby weather stations, or customized forecasts from weather service providers.

> Wind direction: The azimuth direction from which the wind is originating. Direction and strength of prevailing

winds, as well as gusts, are important predictors of fire spread and direction.

State of weather: Numerical descriptors of weather conditions, as well as general observation notes. The following codes are standard in fire weather observation:

0—Clear (cloud coverage less than 10 percent)

1–Scattered (coverage greater than 10 percent, but less than 50 percent)

2–Broken (coverage greater than 50 percent but less than 90 percent)

3–Overcast (coverage greater than 90 percent)

4–Foggy

5–Drizzle (precipitation consisting of numerous fine droplets)

6–Rain

7–Snow or sleet

8–Showers (in sight or occurring at observation point)

9–Thunderstorm in progress (lightning seen or thunder heard, less than 30 miles away)

Cloud formations: Clouds are classified according to elevation and degree of vertical development. Clouds are one indicator of *atmospheric stability*, or the tendency for a hypothetical air parcel to return to its current position after exertion of a vertical force. In general, layered (stratus) clouds indicate a stable environment. Clouds with considerable vertical development are generally indicative of an unstable environment. Another (less obvious) indicator of atmospheric instability is an air mass in which temperatures rapidly decrease with increasing elevation. By contrast, temperatures increasing with elevation indicate an inversion (that is, a stable environment).

Other general observations: Conditions warranting special attention are noted—for example, smoke dispersal, wind shear, presence of dust-devils, and so forth.

Weather observations are collected manually using a portable belt weather kit or they can be sensed via a remote automated weather station (RAWS). RAWS units have sensors that automatically collect, store, and forward weather information hourly via a geostationary operational environmental satellite (GOES), located 22,300 miles (37,167 km) above the equator, to a computer at the National Interagency Fire Center in Boise, Idaho. Weather observations collected at 1300 hours daily on approximately 1,150 RAWS strategically positioned throughout the United States are archived at a federal computer center and can be downloaded off the Internet.

Drought

Weather conditions that increase the threat of forest fires are well known, such as low precipitation and relative humidity, coupled with high temperatures and prolonged solar radiation. High winds complicate matters further. However, the specific meteorological factors are not always known for the climatic periods preceding past fires. Drought, which results from a combination of meteorological factors, is especially problematic (Haines and Sando 1969). Unfortunately the exact role of drought is unknown, probably because the term is used so informally and vaguely.

Unlike discrete events like a wildfire, drought sneaks up quietly, often under the guise of lovely, sunny weather. Meteorologists define drought as simply a shortage of water, usually associated with a deficiency of rainfall, often unpredictable and unanticipated until we are in its midst. We are never sure when a drought begins until it is well under way, and we are often unsure when it ends (McKee et al. 2000). Often in the wildfire community we are made aware of the magnitude of an ongoing drought only by looking backward and examining records for evidence of the lack of moisture or precipitation.

Another problem with describing drought is that there are few universal standards for describing and measuring the phenomenon; that may sound surprising when you consider how great of an impact we seem to attribute to droughts and wildfires. However, drought may have a variety of other impacts—for example, agricultural, meteorological, hydrological, or socioeconomic—and drought will be described and measured differently by each interest group. Furthermore, spatial location is important in defining drought causes and effects. For example, in any given year the 37 cm of average precipitation in northern Colorado may seem bounteous to neighbors in Nevada and Arizona; residents of the eastern United States, however, might view Colorado as being under perpetual drought. So the concept of drought depends on who is affected, the depth of consequences, location, and other descriptors (ibid.).

Predicting drought is somewhat like guessing about the financial markets. Both drought and market runs (to the upside and down) are inevitable, yet no one knows when either process begins, how it will unfold, or how long it will last. Like a bull or bear market, both of which we know will occur in the future, we can be certain that society will need to contend with some form of drought, but we really can't say much more with any certainty.

Amid this confusion in describing drought, especially its impacts on wildfires, numerous measurements have nonetheless been developed to assess the effects of prolonged drying trends on fuels and fire behavior. Some of the more prominent drought measurements of relevance from a fire management standpoint include the following:

Palmer Drought Severity Index (Palmer 1968)

A complicated soil moisture calculation used in the agricultural community to gauge the need for federal drought assistance. It measures the departure of the moisture supply at specific locations from standardized conditions. It has been reconstructed in fire history studies to link fire occurrence with drought periods, including links with El Niño and La Niña cycles.

Keetch-Byram (1968) Index

Assesses fire potential based on the net effect of evapotranspiration and precipitation deficit in deep duff and upper soil layers on a scale from 1 to 800 (Keetch and Byram 1968). It was developed for use in the southern United States but is now mapped nationally.

National Fire Danger Rating System (Deeming et al. 1977)

This system provides perhaps the best known drought measures from a fire standpoint, based on ambient and recent trends in coarse-scale environmental conditions (fuel, weather, topography). For example, the 1,000-hour timelag fuel moisture content (TLFMC) and the Energy Release Component (ERC) have been linked to the occurrence of late-season fire outbreaks in the northern Rocky Mountains.

Normalized Difference Vegetation Index (NDVI) (Burgan and Hartford 1993).

Vegetation Greenness Maps are derived weekly from calculations based on AVHRR satellite observations. Maps include greenness relative to a reference standard, greenness relative to historical ranges for a specific pixel, and departures from average greenness for a particular point in time.

Standardized Precipitation Index (SPI) (McKee et al. 2000)

An index developed in Colorado based on current and historical precipitation data for a particular location, SPI is based on precipitation deviation from the average surplus or deficit for a specific location. It can be used to identify the start and end of drought episodes, if a long enough record is available.

The above list is not intended to be all-inclusive and could be expanded to include other drought indicators in use generally, but nevertheless it illustrates that there is no standard measure of drought that is universally accepted and used by the fire management community—and perhaps there should not be: No single indicator works best for all regions and intended applications, although drought indicators generally may be pretty consistent with one another.

Wildfires and Drought

Unlike a wildfire, drought isn't viewed as a discrete event with a clear beginning and end. Yet drought serves as a precursor to wildfire and other natural disturbances, such as insect attacks on moisture-stressed trees. In fact, some may view wildfire simply as one among several manifestations of a prolonged drought. Even so, drought is sometimes overlooked or underemphasized in terms of causing wildfires, perhaps because of perceptual, definitional, or measurement problems, as noted above. For example, the recent emphasis on fuels abatement in the United States apparently underplays consideration of drought in worsening the effects of recent fire seasons. Perhaps that oversight follows from the difficulty of isolating the relative contributions of fuels versus drought (or their interactions)—at least no one seems able to quantify which is more influential in pushing fire seasons (such as those observed in 2000 and 2002) to the extremes.

What seems clear is that drought does make more fuels available for combustion, and it seems to propel fire behavior beyond thresholds of firefighter control. Drought effects also seem to vary by elevational temperature/moisture gradients. Thus high-elevation, moist-cold ecosystems burn more readily during prolonged droughts than in nondrought years. These systems support plenty of biomass but may not dry sufficiently to burn except during the drought years. Once ignited during a drought, however, these forests burn with characteristic high-severity, stand-replacement fires. At such times firefighters can do little more than attempt indirect attack, work the cooler sectors (for example, flanks and smoldering areas), and wait for a change in weather.

The same principles seem to apply during drought years in lower-elevation, warm-dry forests as well, but for different reasons. Historically, these ecosystems burned more frequently than the high-elevation forests, but decades of fire exclusion also have led to higher surface fuel loads, higher density of smaller diameter trees and understory shrubs (which provide ladders into the tree crowns), and a more continuous fuelbed across the landscape—all contributing to greater likelihood of crown fires. During drought years, these forests also burn uncontrollably.

Certainly the fires that burned in Colorado during the ongoing 2002 drought seemed to follow this trend of uncontrollable fires at both high and low elevations. Thus upper-elevation Engelmann spruce-subalpine fir (*Picea engelmannii-Abies lasciocarpa*) and lower-elevation ponderosa pine–Douglas fir (*Pinus ponderosa-Psuedotsuga menziesii*) forests burned with high severity during 2002. The higher fuel loads and structural changes (that is, increased density of Douglas-fir) in the lower-elevation forests resulting from the decades of fire exclusion may provide a major distinction between the severe burning observed at the two elevations during 2002 in Colorado.

Fire Danger and the Public

The term "fire danger" is used to describe the risk that a fire may threaten the public or cause management problems over a fairly large area, especially as related to antecedent weather and climate. Fire danger is distinguished from the actual or predicted fire behavior resulting from site-specific fuel, weather, and topographic variations. The National Fire Danger Rating System (Deeming et al. 1977) is a complex set of computer programs based on algorithms developed to allow land management agencies to estimate today's or future fire danger for a given rating area, (for example, a ranger district). Indices for the probability of ignition, potential spread, or energy release are developed from standardized (and cumulative) fire weather measurements (such as temperature, relative humidity, precipitation, fuel moisture by size class, wind speed, and direction). Managers rely on archived weather and danger rating indices, along with past and present satellite imagery of vegetation greenness, to assess departures from historical norms—especially useful for tracking the progress of a developing drought. On any given day, the calculations and outputs from the fire danger rating system can be used to develop staffing levels for fire management agencies—or to communicate with the public about potential wildfire problems. Fire managers make use of many communication devices to provide the public with information about fires and fire danger. Audio and visual media, Smokey Bear ads, and visits to local grade-school classrooms are commonly used to inform different audiences. Fire officials will integrate information from numerous sources, such as Red Flag alerts from the National Weather Service, drought indicators, and knowledge of historical trends in fuel availability and fire behavior to increase public awareness of local fire danger.

Perhaps one of the best known devices for communicating with various audiences is the familiar fire danger meter, which can be seen outside most ranger districts and fire stations in forested areas during the summer. Typically, a pointer or different color codes will identify current forest fire danger as low, moderate, high, or extreme. An accompanying image of Smokey Bear may urge forest users to be careful with fire in the forest. Although simplistic in appearance, the rating on any given day may reflect the complexities of fire danger and drought assessment as effectively as any scientific metric.

A complete treatment of fire weather and climate is beyond the scope of this book. Further information on this subject can be found in many references, including Schroeder and Buck (1970) for the United States and Canada, Johnson (1992) for the North American boreal forest, Bradstock et al. (2002) for Australia, and UN Economic Commission for Europe (1982).

Fuels and Fire Behavior

As mentioned previously, the so-called fire environment consists of fuel, weather, and topography (Countryman 1972). Fire managers focus particular attention on fuels because weather and topography are less easily manipulated. Thus fuels management has become a very important fire management strategy for reducing fire hazards in an area. Other management strategies include fire prevention, fire detection, fire presuppression, and fire suppression activities. Each of these activities is discussed generally below (under Fire Protection and Management), although not to the detail found in textbooks on fire (see, for example, Pyne et al. 1996, Chandler et al. 1983a and b, Brown and Davis 1973). Detailed discussion about these activities is beyond the scope of this book, but each needs to be considered while developing fire management plans for an area.

Wildland fuels can be classified into several broad layers or vertical strata, including: (1) ground and litter fuels; (2) woody fuel, lower vegetation, and shrubs; and (3) forest canopy fuels (Ottmar et al. 2002). Each layer contributes uniquely to wildland combustion, although the degree of involvement varies with every fire. Ground and litter fuels include all burnable material (duff, loose surface litter, lichens, and mosses) that accumulates in the open and at the base of trees. Woody fuels include dead, downed twigs and branches, rotten and sound logs, and stumps that might be piled or accumulate in jackpots. Low vegetation includes grasses, sedges, forbs, low shrubs, and tree seedlings. Shrub fuels include taller bushes and needles that might be draped within bushes from overstory tree crowns. Forest canopy fuels include all burnable material in the tree crown (dominant, codominant, and suppressed trees) and dead snags, some of which may form continuous fuel ladders. These layers are not always well defined (as for example with tall shrubs in a forest), but fuel strata provide a convenient framework for thinking about forest fuels.

Ground fuels will burn only when relatively dry. In peat or deep humus, ground fires may smolder for days, consuming fuels in three dimensions at very low rates of spread, say 5 cm/hour. Consumption and mortality of seeds, bulbs, rhizomes, and roots may occur. Local extinctions of species may occur if all reproductive sources are destroyed. Above ground, surface fires generally burn much more rapidly through litter, lower vegetation, and shrub strata. As flames increase in intensity, a litter fire may spread to downed twigs and branches, grasses, and shrubs, and eventually climb ladder fuels and involve tree crowns. Surface fires and ground fires are often linked, though not necessarily. Similarly, surface and crown strata may be linked or burn independently of each other. Thus fire behavior and effects will depend on the characteristics of all affected fuel strata.

Fuel Profiles

A fuel profile describes the ground, surface, and crown attributes that support combustion during a wildland fire. Fuel profiles may reflect historical fire regimes and management practices in a given area, including the characteristic frequency, behavior, and effects of historical fires. The flammability of the area may be described by fuel attributes, such as the loading (for example, tons/acre, tonnes/ha), density, and depth of the fuelbed. Size, moisture, and heat content of fuel particles are among the many additional attributes that affect flammability. In any fire incident, these fuelbed and fuel particle attributes help explain the intensity and severity with which combustion occurs, as well as the effects upon organisms in the area. Even effects downwind or downstream may be explained in part by fire behavior and severity at a particular location.

Geographical location also influences fuel flammability. Some forests appear virtually primed to burn. For instance, eucalypt forests in Australia seem particularly fire prone. In addition to the highly flammable volatiles in the foliage, stringy bark peels off in long strips from individual trees, suspended between surface and live crowns. Bark strips accumulate in curled heaps at the base of trees, creating a porous fuelbed. In southern California, explosive fires periodically rip through the shrub crowns, aided by the volatile chemicals in the foliage and the suspension of dead branches and twigs within the living plant. The needles in stands of eastern U.S. pine barrens literally glisten with volatile

chemicals during prolonged heat episodes, indicating their potential explosive fire behavior. Savannas of Africa, Australia, and
the Americas similarly burn with regularity because of the abundance of fine, flammable fuels.

On any walk in the forest, you may note various fuel,
weather, and topographic features that contribute to fire potential. For example, the following features contribute to the variability of fire behavior in different wildland areas:

> Coarse woody debris (including logging or thinning slash):
> dead, downed woody material, standing snags (dead
> trees), higher fuel loading in larger size classes, spatial
> continuity, the presence of needles (that is, red slash) on
> downed branches from recent management activities such
> as cutting or thinning, or from wind throw.

> Grassland: continuous horizontal and vertical arrangement
> of the fuelbed, high fuel particle surface area to volume
> ratios, and high fuelbed porosity.

> Shrubland: intermediate fuelbed depth, flammability of
> foliage, fine, dead fuels (distributed in the litter and within
> the plant), higher fuel availability when exposed to winds.

> Forest: flammable foliage, fuel jackpots in patches, litter layer
> at the base of trees, compactness of litter layer depending on
> needle length (for example, short- vs. long-needled pines),
> forest canopy that moderates temperature and relative
> humidity extremes, tree stems that reduce wind gusts.

Fires will burn with different signatures through these four fuel
groups. For example, grasslands will burn with rapid rate of spread,
especially when pushed by the winds. High surface-area-to-volume
ratios of fuel particles promote heat transfer and drying. Fire behavior in forests is most complex, with light surface burning when
low wind speeds accompany high moistures; yet those same stands
will support intense crown fires following a prolonged drought. A
fire may have difficulty sustaining itself in a shrub field under cool,
moist conditions in the early morning, yet roar to life and engulf the
shrub crowns later in the afternoon when pushed by winds created
by unequal temperature gradients. Fires burn through areas supporting coarse woody debris with high flame lengths, but also with
prolonged soil heating effects because of extended residence times,
again subject to wind and moisture conditions.

The basic differences in fire behavior among fuel groups are related to the size and shape of fuel particles, the distribution of fuel loads among the various size classes, horizontal and vertical arrangement of fuel particles, and the depth of the fuelbed—as well as wind, moisture, and topographic slope steepness and aspect. The fuelbed and fuel particle differences are represented in standardized "fuel models," after Anderson (1982). These thirteen fuel models essentially become tools for estimating fire behavior (for example, rate of spread and flame length) in different fuel types, subject to environmental influences of wind, fuel moisture, and slope. The thirteen models include:

Grass and Grass-dominated

1. Short grass—Western annual grasses, such as cheatgrass, medusahead, ryegrass, and fescues; live oak savannas; open pine-grasslands.
2. Timber (grass and understory)—Open ponderosa pine stand with annual grass understory; scattered sage within grasslands.
3. Tall grass—Fountaingrass; meadow foxtail in prairie and meadowland; sawgrass prairie and strands.

Chaparral and Shrub Fields

4. Chaparral—Brushfields composed of manzanita and chamise; pocosin shrub field with species like fetterbush, gallberry, and bays; high shrub southern rough with dead limb-wood.
5. Brush—Green, low shrub fields within timber stands or without overstory; regeneration shrublands postfire or other disturbance.
6. Dormant brush, hardwood slash—pinon-juniper with sagebrush; understory mainly sage with some grass intermixed; southern hardwood shrub with pine slash residues; low pocosin shrub field; frost-killed Gambel oak.
7. Southern rough—Southern rough with light to moderate palmetto understory; southern rough with moderate to heavy palmetto-gallberry and other species; slash pine with gallberry, bay, and other species of understory rough.

Timber Litter

8. Closed timber litter—Surface litter fuels in Western hemlock stands; understory of inland Douglas-fir with little fuel; closed stand of birch-aspen with leaf litter compacted.
9. Hardwood litter—White oak litter; loose hardwood litter under stands of oak, hickory, maple and other hardwood species; long-needle forest floor litter in ponderosa pine.
10. Timber (litter and understory)—Old-growth Douglas-fir with heavy ground fuels; mixed conifer stands with dead-down woody fuels.

Slash

11. Light logging slash—Slash residues left after skyline logging; mixed conifer partial cut slash residues (similar to closed timber with down woody fuels); light logging residues with patchy distribution.
12. Medium logging slash—East-side ponderosa pine clearcut; cedar-hemlock partial cut; lodgepole pine thinning slash.
13. Heavy logging slash—West Coast Douglas fir clearcut with high quantity of cull; highly productive cedar-fir stand with large quantities of slash.

The thirteen standard fuel models listed above do not cover all fuel profiles encountered in the United States, but the models actually cover a surprisingly large number of ecosystems and wildland environments. Wise practitioners use the models to provide guidance in making management decisions, fully aware of their limitations. In particular, they will not fixate on the names of fuel models, but look instead at how well a particular model reflects local conditions. For example, a logging slash model (#11–13) may adequately represent a forested area that has experienced an extreme wind storm event or a beetle infestation with subsequent fall-down of dead trees—that is, conditions not the result of human activities such as logging or thinning.

With practice, custom fuel models can be constructed using inventoried information for other areas not adequately represented by standard fuel models. In fact, practitioners with an adequate modeling background and good fuel inventory data can build and use many more fuel models, especially when calibrated with actual fire behavior observations.

Obviously, many factors contribute to fire spread and intensity within a fuel complex. A trained eye is able to understand fire behavior by focusing on vegetation as fuel, while melding the local influences of weather and topography.

Standard Fuel Measurements

Because of their importance in influencing fire behavior and effects, several methods have been developed for estimating fuel loading and depth in the field. The most accurate, but time-consuming, methods involve measurement techniques such as the planar intercept technique (see, for example, Brown 1974) for coarse woody debris, or clipping and weighing of plant and litter materials. Fuel inventory techniques (see, for example, Brown et al. 1982) can be time consuming but provide the practitioner with a sense for the distribution and variability of fuels in wildland ecosystems.

As an alternative to actual fuel measurements, photo series have been developed to provide managers with quick estimates based on inventories paired with photographs. Photo series provide estimates for the following fuel attributes (Ottmar et al. 2002, with units converted to metric):

Woody loading by size and rot class (tonnes/ha)

Litter/duff/moss/lichen depth (m) and bulk density (kg/m^3)

Mineral soil exposure/soil type (percent)

Fuel depth (m)

Herb and shrub loading (metric tons/ha)/height (m) by dominant and codominant classes

Tree species/diameter (cm)/height (m) by dominant and codominant classes

Live and dead stems per hectare by diameter class

Seedling density (#/ha)

Canopy layers, canopy cover (percent), height to crown base (m)

The thirteen standard fuel models most likely will underestimate total fuel in an area, since they focus primarily on surface fuels that are available to burn. Thus, fuel inventory information may not be suitable for insertion in standard fire behavior models without calibration and testing against fire behavior observations in the field.

In the future, additional measurement schemes are likely to arise. For example, fuel characteristic classes are under development that more accurately capture the structural complexity and geographical diversity of fuelbeds. More accurate fuel description is required as more sophisticated models for predicting fire behavior and effects and for making large-scale landscape assessments are developed. (http://www.fs.fed.us/pnw/fera/fccs/index.html).

Fire Behavior Prediction

As noted in Table 5–3, stages in the life of a fire include preheating, flaming, glowing and smoldering, then cooling and extinction. From its onset, numerous fluctuations may occur in fire growth, including periods of acceleration in rate of spread, and irregular pulses in active flaming that precede eventual extinction at various points along the fire perimeter. Similarly, gyrations in fire activity can be expected as the fire responds to moisture, wind, and topographic slope gradients. Rothermel (1983) details methods for predicting fire behavior based on fuel, weather, and topographic measurements, assuming a steady-state fire growth rate. His model of fire spread, based on basic principles of thermodynamics and laboratory experiments, is briefly summarized here because it has been widely accepted and applied by fire managers.

The first law of thermodynamics states that energy is neither lost nor gained in all energy processes, including combustion, while the second law implies that energy conversion processes result in nonusable energy (Joesten et al. 1993). Rothermel (1972) relied on these principles to develop a fire spread equation (4) that suppression crews could use to predict the rate of spread (R) in wildland fuels:

$$R = \frac{I_R \, \zeta \, (1 + \phi_w + \phi_s)}{\rho_b \, \varepsilon \, Q_{ig}} \qquad (4)$$

where
ρ_b is fuelbed bulk density, the amount of fuel occupying a unit volume of the fuel array (lb/ft^3);
ε is the effective heating number, a measure of the fraction of the potential fuel that must be raised to ignition (unitless);
Q_{ig} is the heat of preignition, the energy per unit mass required for ignition (Btu/lb);
ϕ_s is the slope factor, accounts for improved radiation, convection, and air flow because of slope increase (dimensionless);
ϕ_w is the wind factor, representing a dimensionless multiplier that accounts for the increased rate of spread resulting from improved radiant and convective heat transfer and oxygen flow accompanying a wind-driven fire (dimensionless);
ζ, the propagating flux ratio, is a dimensionless number that indicates the proportion of the reaction intensity that preheats the adjacent fuel particles;
I_R is the reaction intensity, a measure of the total energy release rate per unit area of fire front ($Btu/ft^2/min$).

Essentially, equation 4 expresses rate of spread as a ratio between two quantities: the heat generated to propagate a fire divided by the heat required to ignite fuels ahead of the fire. The bigger that ratio, the faster the fire will spread. Rothermel (ibid.) developed relationships between fuel particle and fuelbed properties that could be measured in the field or represented by standardized fuel models (Anderson 1982) to calculate the terms in (4) and thereby produce fire behavior estimates, given estimates for midflame wind speed, fuel moisture, and terrain slope. This model is the foundation for fire behavior prediction systems in the United States, including BEHAVE (Andrews 1986), BehavePlus (Bevins 2003), and FARSITE (Finney 1998). The model also has been reformulated using metric equivalents (Wilson 1980) and programmed for a handheld personal digital assistant (PDA).

Once rate of spread (R) is calculated by equation (4), additional fire estimates can be calculated based on fuel particle and fuelbed descriptors, specific to each fuel model. These include the following calculations (from Andrews 1986):

$$t_r = 384/\sigma \tag{5}$$

$$D = R\, t_r \tag{6}$$

$$H_A = I_R\, t_r \tag{7}$$

$$I_B = I_R\, D/60 \tag{8}$$

$$F_L = 0.45\, I_B^{0.46} \tag{9}$$

where:
t_r = flame residence time (min);
σ = characteristic surface-area-to-volume ratio (ft^{-1});
D = flame depth (ft);
R = rate of spread (ft/min), from (4);
H_A = heat per unit (Btu/ft^2);
I_R = reaction intensity (Btu/ft^2/min);
I_B = Byram's fireline intensity (Btu/ft/sec);
F_L = flame length (ft).

Thus equations 4–9 make possible the calculation of fire characteristics, such as surface fire rate of spread (ft/sec, ch/hr, or m/sec), fireline intensity (Btu/ft-sec or kW/m), flame length (ft or m), which can be used to provide guidance for fire suppression activities. Residence time (sec) provides limited information for predicting fire effects; additional crude inferences can be derived regarding crown scorch and tree mortality, based on extant studies. Rate of spread estimates also are used to develop projections for the fire's area (acres or ha), and perimeter (ft, ch, or m) at user-specified time intervals (hr). These size estimates presume that a free-burning fire burns as an ellipse. It follows that fire shape can be described roughly by the major and minor axes of an ellipse, or the length-to-width ratio.

Most of these descriptors (that is, in equations 4–9) are readily observable in the field, notably rate of spread and flame length. Other descriptors may not be so easily observed—for example, heat per unit area or reaction intensity. Even so, certain types of fire can be characterized rather simply through tools such as the fire characteristics chart (Figure 5–12). Note how suppression difficulty can be inferred with an estimate for rate of spread and one of the following: flame length, fireline intensity, or heat per unit area. The chart also delineates thresholds for fire control efforts that can be used by firefighters. Thus hand crews should be able to suppress a fire with flame lengths less than 4 ft (1.2 m), whereas bulldozers might be required for fires with larger flames. Suppression may be difficult if flame lengths exceed 8 ft (2.4 m) because of tree torching, and impossible when flame lengths exceed 12 ft (3.6 m).

Numerous computer programs (see Chapter 7) have been developed to aid decision-makers, based on Rothermel (1972) with subsequent modifications. In addition to the computerized models (for example, BehavePlus and FARSITE) mentioned above, other notable example tools include NEXUS (a spreadsheet linking surface with crown fire behavior) and Flammap (a planning tool for tracing historical burn patterns on the landscape). Equation 4 also serves as the basis for computing fire danger ratings and a helpful tool known as a pocket card (from a useful program called FireFamily Plus), which provides incoming firefighters with general guidance regarding expected local fire behavior. A different approach is used by a program called BURNUP to estimate burning characteristics after the flaming front has passed— that is, during glowing and smoldering combustion. Lastly, the computer program FOFEM (for First Order Fire Effects) calculates short-term fire effects (such as on plants and soils), using a fuel modeling approach that differs considerably from the thirteen fire behavior fuel models from Anderson (1982).

The U.S. fire behavior prediction system has been adapted for use elsewhere, such as in Mexico and Europe. By contrast, other countries such as Canada and Australia have taken different approaches, based in part on the different fuel complexes present elsewhere but also the result of different modeling philosophies. For example, the Canadian and Australian fire behavior prediction systems rely heavily on empirical relationships developed from fire observations in the field. Managers in Alaska sometimes use the Canadian system because of similarity in fire environments.

Extreme Fire Behavior

Uncontrollable fires stem largely from conditions in the fire environment that promote extreme fire behavior. Extreme burning conditions promote crown fires and ember-spotting, both of which can lead to major fire runs and threats to firefighter and public safety. The comprehensive study of extreme fire behavior is beyond the scope of this book, but we will focus here on two

Figure 5-12
Fire Characteristics Chart
The fire characteristics chart (Andrews 1982) allows firefighters to
relate fire behavior predictions and observations to the difficulty of
suppressing fires. For example, Fires A and B have the same flame
length and fireline intensity, but dramatically different spread rates
and heat per unit area.

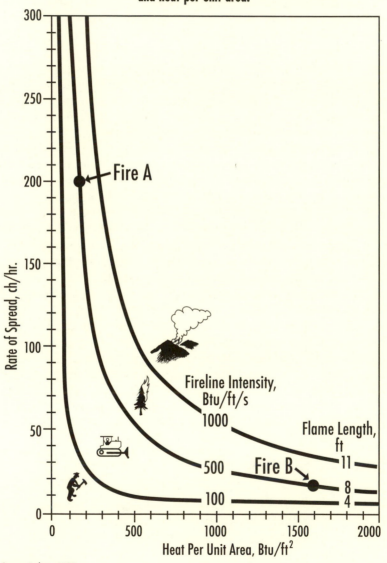

Source: Andrews 1986

particularly problematic types: crown fires and spot fires. Both terms indicate erratic or severe burning conditions that dramatically change the face and effects of a fire but can also create spectacular threats to firefighters. Commonly there are many situations that lead to extreme fire behavior, but major fire runs are generally triggered by the alignment of wind speed and direction, topographic slope, and conducive fuel characteristics (such as moisture and temperature). Under those circumstances crown fires and spot fires become more likely.

There are three general crown fire phases: active, independent, and passive. Active crown fires occur where fire on the surface is linked to fire spread in tree crowns over an extensive area. Independent crown fires are characterized by fire spread through the tree crowns, largely uncoupled from fire at the surface. Passive crown fires occur when individual trees torch because of vertical heat transfer from surface to crown. Foliage on adjacent trees is sparse or otherwise unavailable. Crown fires figure prominently in most of the megafires listed in Chapter 3, whether involving a forest or shrubland canopy. Plant and stand attributes con- tributing to crown fire potential include the following:

1. Low height of tree crown bases (that is, the forest canopy base starts low to the ground);
2. Presence of fuel "ladders" such as dead/live twigs and leaves, dry bark, and dead shoots on tree trunks—or understory trees that reach into the forest canopy;
3. Low live fuel moisture (less than, say, 80 to 100 percent);
4. Low mineral concentrations in fuels;
5. High concentration of volatile oils and resins in foliage and leaves/needles;
6. High surface area to volume ratios of leaves, needles, and twigs;
7. Abundant litter fall with low decomposition rates;
8. Relatively dense forest canopy;
9. Flammable understory strata, including shrubs, low vegetation, tree seedlings, and coarse woody debris;
10. Presence of dwarf-mistletoe brooms, other insect/pathogen infestations, or mosses or lichens suspended within the tree that enhance crown flammability.

Spot fires can be started when live fire brands are ejected above or ahead of the flaming front, are entrained in transport winds, and then start new fires as they land in receptive fuelbeds in advance of the main body. Many natural fuels make suitable firebrands: cone scales, grass clumps, bark flakes, twigs and other branchwood parts, and moss. If midflame windspeeds exceed 10 mph and fine fuel moisture is less than 10 percent, spotting can occur whenever fireline intensities exceed 700 Btu/ft/sec. However, if winds become stronger, the likelihood of spot fires can actually decrease, as embers can burn out during transport, landing harmlessly downwind. Even so, spot fire distances of 20 km (12 mi) ahead of the main fire front have been documented in Australian eucalypt forests.

Windborne embers that land in the tops of trees can spread a crown fire. Embers that land in the tops of tall dead trees or snags can be especially problematic for firefighters, as the fire may continue to burn and then smolder for days or weeks if left unattended. Leaving the snag to burn out may be an option if it is located well inside the fire perimeter. However, a snag burning close to fire control lines may need to be felled, in order to reduce the likelihood of future spot fires or of the burning snag's falling over and starting fires outside the burn perimeter. Moreover, a snag is inherently hazardous to work around, especially if it must be felled to squelch fire at the top. Trees and snags with rotted wood in the bole or branches can be extremely hazardous to cut down, with broken wood or falling branches (called "widowmakers" by old timers) likely, or an unpredictable fall always a possibility. What's more, snags can be valuable habitat for cavity-nesting birds and therefore desirable in areas being managed for certain favored species. Similar problems arise if lightning causes a fire in the top of a tall live tree, although live trees are inherently easier to cut down because the wood is more solid and the felling direction more predictable. Even so, keeping the felled tree or snag within existing fire control lines may be problematic.

If a fire develops a major run or shows extreme convective development, it is sometimes called a blowup fire. A rapid blowup is sometimes called a firestorm. Very large fires of high intensity are known as mass fires. The explanatory contributors to blowup fires are poorly understood, although the "alignment" or confluence of low fuel moisture, winds, intense preheating of fuels from solar radiation, slope, and the fire itself are important in organizing major fire runs (Campbell 1991). Fires with very high rates of spread are known as conflagrations, or sometimes simply as "wind-driven" fires.

Large fires that burn under the influence of a massive convection column are also called "plume dominated." Generally, a billowing plume indicates simultaneous ignition of fuels covering a fairly large area. Well-formed convection columns develop under unstable atmospheric conditions, and they can dominate a fire's behavior with updrafts and downdrafts (much like a thunder cell). A particularly dangerous situation arises when the plume collapses on account of gravity—the strong downdraft is capable of pushing a fire in any direction. Fire behavior specialists in the 2002 Rodeo-Chediski fire noted several collapsed plumes during major runs. A collapsed plume was implicated in the six fatalities during the 1990 Dude fire.

Albini (1979, 1983) developed the theory behind calculations for the maximum downwind spot fire distance that has been incorporated into most fire behavior prediction systems. His theory predicts intermediate-range spotting, at a greater distance downwind than debris blowing across a control line, for example. Firebrands are assumed to be small enough to be lofted and carried down wind, but large enough to start a fire when deposited on the ground. The actual onset of a spot fire ignition depends on the receptivity of fine fuels where embers are deposited—that is, the probability of ignition, which depends on temperature and humidity.

Rothermel (1991) used the notion of plume versus wind driven large fires to develop his crown fire nomograms, or sequential charts, for developing large fire spread predictions. Originally developed by Byram (1959), the theory relies on a calculation of the power of the fire, or rate of energy released, calculated on a per area basis, as a reflection of the energy source producing the convection column. The heat energy produced by the fire produces the convection column because of the temperature rise of the air, which lowers its density, producing the buoyant vertical force. This force is compared with the rate of flow of kinetic energy in the wind field, expressed in the same units. Whichever is greater (power of fire or power of wind) will dominate the fire's spread.

The Haines Index (Werth and Werth 1998) provides useful guidance for predicting large wildfire growth and extreme fire behavior, based on atmospheric stability and moisture content of the lower atmosphere. In particular, the index looks at temperature differences at two different elevations (approximately 3,050 m and 5,500 m), and the temperature-dewpoint difference at 3,050 m, to calculate an index from 2 to 6. Potential for extreme fire behavior is greatest with a Haines Index of 5 or 6, including high spread rates, extensive spotting, active crowning, and the development of large convection columns.

Fire managers thus have several tools, both automated and not, for assessing extreme fire behavior. A fire behavior specialist will likely rely on the index that seems to provide the most reliable predictions in the local area, combined with knowledge and intuition informed from previous fire experiences.

Historical Fire Regimes

Recurrent fires have played an important, and sometimes critical, role in the evolution and development of many wildland ecosystems worldwide. Most ecosystems bear unmistakable imprints of previous disturbances (fires, floods, windstorms, avalanches, and so forth). The evidence for historical fires can be seen in the mosaic of vegetation covering a landscape, which may consist of plants with various adaptations to fire, as well as the structure (for example, size, age, stem diameter distribution, species composition) of patches within the landscape. Tree stems may bear scars from fires that can be dated, especially in forests characterized by frequent, low-severity surface fires. In other areas—for example, higher-elevation forests that support infrequent, crown fires—the evidence for past fires may be evidenced from the even-aged cohorts of trees that regenerate postfire. In savannas (grasslands with intermittent trees and shrubs), tree density will increase in the absence of fires and decrease where fires are allowed to run their course.

Still other evidence has been extracted from pollen and charcoal fragments from lake or swamp sediments or from peat bogs. Other signatures may be more difficult to identify, such as patches in the landscape that have experienced a mixture of high- and low-severity fires. In some cases, indirect evidence for fire frequency may be reflected in the characteristics of plants occupying a particular site. Although speculative, the clumpiness of patches or the age structures of individual patches may shed considerable light on historical fire habitats. These various sources of evidence have been used to establish the frequencies with which fires have burned in various ecosystems in the past, especially when formal (that is, public agency) records extend back only to the mid-twentieth century. Fossilized charcoal, lake sediments, and tree fire scars are common indicators that can extend the fire record several centuries previous to agency fire reports. Using such evidence, the historical fire return interval (the number of years between successive fire events) has been summarized for a wide variety of environments (Table 5–5).

Table 5-5
Representative Fire Return Intervals (in years) from a Variety of Biomes/Ecosystems Worldwide, Illustrating the Wide Range of Fire Recurrence in Different Vegetation Types

Biome/ecosystem	Location	Return interval (yr)
Tundra		
Tundra	Canada, Alaska	500
Alpine tundra	New England	1000+
Boreal forest		
Open *Picea* forest and lichens	Alaska, Yukon	130
Picea glauca forest, flood plains	Alaska, Yukon	200+
Pinus banksiana forest	Northwest Territories	25-100
Pinus contorta forest	British Columbia	50
Subalpine forest		
Pinus contorta forest	Montana	25-150
	Sierra Nevada, California	100-300
Abies balsamea forest	New England	1000+
Moist temperate forest		
Pinus palustris/Adropogon forest	SE United States	3
Nothofagus forest	Tasmania	300
Eucalyptus forest	Tasmania	100
Grasslands		
Grasslands	Tasmania	10-25
Prairie	Missouri	1
Swamps and marshes	SE United States	30-100
Dry temperate forest		
Mixed conifer forest	California	7-10
Mixed conifer/*Sequoia* forest	California	10-100
Mediterranean-climate vegetation		
Evergreen chaparral	California	20-50
Deciduous chaparral	California	30-100
Semi-arid deserts		
Desert-scrub	Arizona	50-100
Pinus/Juniperous woodland	W United States	100-300
Tropical vegetation		
Moist evergreen scrub	Florida	20-30
Tropical forest	Equatorial areas	Unknown

Source: After Whelan 1995 and Chandler et al. 1983a.

Although interesting in its own right, the fire recurrence intervals in Table 5–5 indicate little about the types of fires or their ecological effects. Generally speaking, longer intervals are found in wet-cold or dry-hot environments, where ignitions are rare on account of fuel unavailability. Thus long return intervals are found in alpine and subalpine forest as well as in arid deserts. The shortest fire return intervals are found in grasslands and dry temperate forests, productive ecosystems that dry out sufficiently to burn quite frequently.

The notion of a fire regime (discussed below) provides a useful construct for organizing information about long-term fire trends and prominent fire effects in specific ecosystems, at least at a coarse scale of resolution. Although admittedly artificial, the fire regime concept brings a degree of order to a complicated body of fire behavior and fire ecology knowledge, including recognition that fires produce unique effects in different ecosystems. Fire regimes may be characterized by various descriptors, but they usually include frequency, intensity, and seasonality of historical fire events. Additional descriptors, such as areal extent and patchiness of burning, also may be noted if data are available.

Fire frequency (# fires/time) or its useful analog, the mean return interval (yrs) between successive fires, can be estimated by dating fire scars on trees or analyzing age classes of current vegetation. Fire intensity, or rate of energy release, is commonly used to describe the type and heat output from the fire event. The term "fire intensity" is often used interchangeably with the term "fire severity" to describe the effects of fire on an ecosystem, though severity more accurately reflects the influx of heat received by aboveground vegetation and belowground soil resources (Ryan and Noste 1985). Both descriptors have been shown to influence the ability of different plant species to withstand or regenerate after a fire. For purposes of this discussion, distinctive categories of fire intensity/severity include surface, understory, and crown fire events.

Fires typically burn in characteristic time periods or seasons of the year, when ignitions are likely to occur in fuels that will support combustion. This usually occurs during warm-dry periods (such as the summer months). Summer lightning associated with convective storms provides a common ignition source in North America, occurring at a time when fuels are desiccated because of prolonged solar radiation. Historically, fires ignited by lightning in early summer may have burned for prolonged periods, sporadically subsiding because of rainfall but reigniting following prolonged drying or in response to high winds (such as foehns). By contrast, fire season peaks in southern California during Fall (because of foehn winds) and in Florida during January (dryness). Arizona and New Mexico may experience a split fire season, with peak activity in late Spring followed by a secondary increase in fire activity in the Fall, with the moist monsoon reducing fire occurrence in between. In some places, such as Colorado, fuels are dry enough to support combustion in January or February although peak burning conditions may occur in late Spring and Fall.

Reconstruction of the areal extent of historical fires is difficult. The dating of stand ages across the landscape may provide some insight, however, and cross-dated fire scars may provide hints in areas frequented by low-intensity fires. The areal extent of fires in ecosystems that evolved with rare, high-intensity fires may be revealed through the pattern of regeneration following the last fire event. This pattern usually can be observed on forested hillsides, where abrupt changes in tree size classes or species will create visible patchiness in the vegetation mosaic— evidence that may persist for decades or centuries. Each of these methods has shortcomings that affect the accuracy of estimates and their usefulness for making management decisions. For example, unburned patches will be found in even the most widespread and severe fires. Thus the landscape provides a historical tale for the attentive observer.

The evidence for past severity is most reliably estimated for areas that burn frequently with insufficient severity to kill standing trees. Large trees in such areas develop adaptations to withstand the characteristically low-intensity fires, such as thick bark that protects the live cambium. Fires that burn at the base of a tree may kill the vascular cambium without killing the plant and may leave distinctive lesions (or fire scars) that can be cross-dated using dendrochronological methods for dating tree rings. Accordingly, tree rings and fire scars can be cross-dated with a known master chronology for an area, usually constructed from ten to twenty mature, undisturbed trees in the local area (Agee 1993). Such a comparison usually can help identify missing or false rings in the sample trees. Thus repeated, low-severity burning of individual trees provides a historical record of fire recurrence (Brown and Shepperd 2001). When pooled, the fire scar records from multiple trees can be used to develop a composite fire interval for an area (Agee 1993; Stokes and Dieterich 1980).

These techniques not only allow researchers to develop estimates for the frequency with which fire historically has spread through an area but also may suggest changes in fire regime because of climate fluctuations or management policies. Also, spring versus late summer ignitions may be identifiable in the fire scar record, based on a scar's placement within the tree's annual growth ring—possibly indicating the severity of historical fires in ecosystems where late summer fires follow prolonged drying and burn more severely than fires occurring earlier in the year.

The extent or size of past fires can be roughly estimated if fire scars are dated from multiple trees sampled from a large enough area, though these areal estimates may be uncertain. The scar location within a tree ring can also hint at the season of burn, although intensity or severity of burning is more difficult to assess.

Fire-scar dating techniques also have been used to identify changes in fire frequency associated with different land uses or management policy changes. One of the most studied changes is the precipitous drop-off in fire activity accompanying the institution of fire exclusionary practices in the twentieth century (see, for example, Brown et al. 1999 for Colorado; Swetnam and Betancourt 1990 for the U.S. Southwest; et al.). In fairness, a comprehensive analysis of past fires should focus on all human impacts, including livestock grazing and timber harvesting, that could impact fire frequency, size, and severity.

Heinselman (1978) was one of the first to formalize the idea that historically characteristic fire frequencies and fire types could be associated with different ecosystems. He identified seven major fire regimes, ranging from areas with no, or little natural fire occurrence to areas with frequent, low-intensity fires, as well as areas with infrequent, high- and low-intensity fires. For example, in the western United States, very warm systems (such as deserts) or very cold systems (such as alpine tundra) may have no or little natural fire occurrence. Sierra Nevada and Southwest ponderosa pine (*Pinus ponderosa*); giant sequoia (*Sequoiadendron gigantea*)–mixed conifer forests in California, and some native grasslands burned with frequent low-intensity surface fires every 1 to 25 years. By contrast, high-elevation spruce-fir (*Picea engelmannii–Abies lasiocarpa*) forests in the central Rocky Mountains, cedar-hemlock (*Thuja plicata–Tsuga heterophylla*) in the Cascades, and grand-fir (*Abies grandis*) in the northern Rockies may burn as high-severity stand replacement fires every 300 to 800 years or more. In between those extremes, Rocky Mountain ponderosa pine burned with infrequent low-intensity surface fires every 25 years or more, while Coast redwood (*Sequoia sempervirens*) burned with infrequent, high-intensity surface fires over a similar period. Southern California chaparral, aspen (*Populus tremuloides*), pinon-juniper woodlands (*Pinus edulis-Juniperous spp.*), and sagebrush (*Artemesia tridentata*) grasslands characteristically experience relatively moderate return intervals (25 to 100 years), with stand-destroying crown fires. Lodgepole pine (*P. contorta*); Western white pine (*Pinus monticola*); Western larch (*Larix occidentalis*); and Douglas-fir (*Psuedotsuga menziesii*) are among the ecosystems that experience variable fire regimes—that is, frequent low-intensity surface fires as well as high-intensity crown fires after a long return interval (100 to 300 yr).

Subsequent authors have taken issue with the number of regimes classified above, usually opting for a lower number. Thus Kilgore (1981) proposed six different fire regimes, while Hardy et al. (1998) and Morgan et al. (1998) proposed five. Brown (2000, p. 5) synthesized the differences between those various classification schemes and arrived at a suggested categorization based on four regimes, with no mention of frequency of occurrence:

1. Understory fire regime (applies to forests and woodlands)—Fires are generally nonlethal to the dominant vegetation and do not substantially change the structure of the dominant vegetation. Approximately 80 percent or more of the aboveground dominant vegetation survives fires.
2. Stand-replacement fire regime (applies to forests, woodlands, shrublands, and grasslands)—Fires kill aboveground parts of the dominant vegetation, changing the aboveground structure substantially. Approximately 80 percent or more of the aboveground dominant vegetation is either consumed or dies as a result of fires.
3. Mixed severity fire regime (applies to forests and woodlands)—Severity of fire either causes selective mortality in dominant vegetation, depending on the susceptibility to fire of different tree species, or varies between understory and stand replacement.
4. Nonfire regime—Little or no occurrence of natural fire.

Not surprisingly, opinions vary on the appropriate number of fire regimes that should be used to characterize fire activity across continental scales. In reality, the differences between the various schemes are inconsequential compared with the main point that fire occupies different niches in various ecosystems. This irrefutable point is more important than quibbling over artificial distinctions about the actual number of fire regimes. Thus the above regimes and their descriptors should not be viewed as rigorous classifications, but rather as a convenient (and artificial) way of viewing fire in natural ecosystems. Furthermore, the appropriate number of distinct fire regimes often depends on the intended scope of a particular investigation.

For example, Schmidt et al. (2002) maps (on a coarse scale) the continental United States into five fire regimes characterized by different historical fire frequencies and severities, wherein fire frequency is the number of years between successive fires and fire severity is defined by the effect of fire on overstory vegetation (grass, shrub, or tree); it is also available at http://www.fs.fed.us/fire/fuelman/firereg2000/maps/fr2000.pdf. In low-severity fires, more than 70 percent of the basal area (m²/ac) or more than 90 percent of the canopy cover of the overstory vegetation survives (Morgan et al. 1996). Mixed-severity fires produce moderate effects in the overstory, cause mixed mortality, and produce irregular spatial mosaics resulting from different fire severities (Smith and Fischer 1997). Stand replacement fires consume or kill more that 80 percent of the basal area or more than 90 percent of the overstory canopy cover (Morgan et al. 1996).

The frequency and severity of recurrent fires may reflect the moisture and temperature gradients that characterize an area. For example, in the Pacific Northwest, Agee (1993) differentiates low-, moderate-, and high-severity fire regimes based on temperature and moisture stress indices associated with dominant vegetation types. For instance, mixed conifer ecosystems with ponderosa pine (*Pinus ponderosa*) in high temperature–high moisture stress areas most likely were established in the presence of frequent (nonlethal) fires that burned at 5 to 15 year recurrence intervals. Red fir (*Abies magnifica*) stands, found in low temperature–low moisture stress environments, burned with moderate severity every twenty-five to seventy-five years. Western hemlock–Douglas-fir (*Tsuga heterophylla-Psuedotsuga menziesii*) stands, growing mostly in high temperature–low moisture stress areas, most likely were established following infrequent (that is, greater than 100 year recurrence) lethal severity burning that resulted in stand replacement.

Understanding the historical role of fire provides insight into long-term sustainability, plant adaptations and current patterns in the vegetation mosaic, as well as guidance for future management activities. For example, fire managers might choose to restore fire in areas where fire has been excluded because of fire suppression activities, using intentional prescribed ignitions. Furthermore, characteristics such as tree resistance (the result of bark thickness or branching habit), plant reproductive strategies (seed versus sprouting from root crown), or even flammability may affect fire recurrence. The distribution of even-aged clumps within a heterogeneous landscape may reflect historical fire patterns. Managers use these and other clues to unravel stand and landscape fire histories and to plan best strategies for meeting land-use objectives.

Unfortunately, our understanding of fire regime descriptors for most North American ecosystems is limited and sometimes subject to misinterpretation. Additionally, fire is but one disturbance agent with which ecosystems have evolved. Frequent low-intensity surface burns may reduce bark beetle outbreaks by thinning stands and reducing competition for available resources among remaining trees. By contrast, fire-damaged trees may attract beetles and other insects to a burned area. Smoke inhibits some fungi and dwarf mistletoe, but it may lower air quality in the short term. In addition, fire injuries to living tissue may enhance the onset of other decay mechanisms.

Human activities such as harvesting, grazing, and fire exclusion have altered ecosystems to varying degrees, with consequences most pronounced at lower elevations. Similarly, decisions to withhold management from an area may lead to unintended consequences. For example, an unthinned or uncut forest may provide prime breeding grounds for bark beetles that attack less vigorous, stagnating stands. Always present at endemic levels, a mountain pine beetle (*Dendroctonous spp.*) epidemic leads to increased levels of fire hazard as branches and needles fall from dying and dead trees. A second pulse in fire hazard may become evident within a decade as dead trees (snags) fall over and accumulate on the forest floor. Fuel levels in such areas may require restoration treatments in order to meet land management objectives and improve forest health. These treatments may include mechanical, biological, and chemical options, or other strategies, including prescribed fire.

In the past, disturbance impacts on ecosystem structure and function were considered as external in origin. Only recently have the effects of fires and other disturbances been considered as originating from within, or being characteristic to, a particular plant community. Like other disturbances, fires may vary in intensity from benign to catastrophic, depending on a myriad of environmental influences. Fuel buildups and insect/disease infestations may also increase the severity of fire events. Although all disturbances can be disruptive, the effects of fire may be modified through judicious management of the fuels complex. In fact, unlike other disturbances, fire itself may be used to lessen or manage the effects of future events—that is, through careful application of prescribed fire to achieve management objectives.

In summary, the notion of a fire regime is useful and synthetic, but it must be interpreted with caution, especially inasmuch as estimates for fire return interval and historical severity are constructed from indirect data sources and subject to high degrees of variability. The extension of inferences about fire frequency to areas of similar vegetation type or plant species may not be warranted—for example, ponderosa pine forests in the Southwest have been subjected to dramatically different fire regimes than have the same species growing in the Rocky Mountains of Northern Colorado or in the Sierra Nevada of California.

Fire Effects

Fire effects are among the most fundamental considerations driving fire policy development and implementation. Besides intense heat, a fire generates particulates, water vapor, and other gases—all of which are added to the atmosphere, at least temporarily. Removal of the soil organic layer can greatly reduce site productivity, as well as producing changes in soil temperature and moisture profiles. Downstream or off-site consequences may result from flooding or soil erosion. The potential for fire to cause resultant damage is the sole reason that we have fire suppression policies. At the same time, fire's beneficial effects are the basis for fire use programs. Fire management involves the striking of a balance between fire control and use—that is, suppressing fire to prevent damage versus using fire or allowing it to promote ecosystem processes.

Most of fire's direct effects on vegetation are the result of the transmission of heat (see equations 1 and 3). Heat from open fires passes into the atmosphere through convection (about 75 percent) and radiation (roughly 25 percent). Relatively small amounts of heat pass into the soil substratum, through a combination of convection, radiation, conduction, and direct flame contact. This small fraction of heat pulse to the soil becomes increasingly important as fire duration increases. For example, roots, cambium, and soil micro-organisms can be killed or damaged by prolonged exposure to heat at the base of a tree.

The amount of heat received by vegetation and flammable materials depends on the energy transmission processes involved. Radiation barely penetrates the surface of an unburned fuel particle. The depth of heat penetration into exposed plant parts depends on the absorptivity of the receptor. Smoke reduces energy transmission through the atmosphere. Convection surrounds plant parts with heated air, which enters the receptor through conduction. Plants respond to heat with desiccation and heat injury. Wind reduces the effect of convective heating by dispersal. Heat injury results from the time/temperature profiles. The protoplasm of most plants is killed by exposure to 50° C (122° F) for longer than ten minutes. At 60° C (140° F), death can be instantaneous. Thus plant death can occur via short exposures to high temperatures or longer exposures to lower temperatures (Miller 2000). Seeds, bacteria, fungi, and other organisms in the soil may be shielded from initial injury, but they might succumb from prolonged heat exposure.

Fire effects will depend on fire environment and the type of fire behavior (Ryan 2001). A fire's effects on plants may vary considerably between different fires and even within the same fire, just as a fire behaves differently depending on location (head, rear, or flank) and interactions with its environment. Thus a surface fire may scorch needles and kill individual trees, whereas a crown fire may leave behind a stand of dead trees or snags. Fire behavior, heat duration, the spatial pattern of fuel consumption, and the distribution of heat below the soil surface all influence injury and mortality of plants, as well as their subsequent recovery (Miller 2000). Postfire responses of plants to fire also depend on plant growth characteristics, susceptibility to heat, and recovery mechanisms. Since fires may occur at any stage of a plant's development (growth, flowering, fruiting, or dormancy), a variety of adaptive traits may allow an individual plant to persist in a fire environment.

Sensitivity to heat is determined by the characteristics of plants as well. Plant mass, thickness, density, and moisture content will affect heat sensitivity, as well as the location of heat-sensitive tissues such as live buds or stems. These tissues are much more sensitive to heat when actively growing and with high moisture contents (Wright and Bailey 1982). Thicker bark provides better insulation to cambium, the live growing tissue located just beneath a tree's bark. Similar protection is afforded to sensitive tissue by bud scales or insulating soil layers that protect roots. Also, the lower a plant's initial temperature, the longer or more intense the heating required to produce lethal internal temperatures. Succulent young shoots are more susceptible to damage than older plants; seeds and dormant buds are more resistant. Different species of tree thus exhibit varying degrees of resistance to fire injury, based on bark thickness, branching habit, and the insulation of vital growth parts (Table 5–6).

A large tree with low-density branching pattern and taller height to live crown will likely have a better chance of surviving a surface fire than will smaller plants with a dense crown extending to the surface; smaller plants are more likely to die or be injured in a fire, as they will be closer to the flames and more susceptible to crown scorching, foliage consumption, and damage to tree boles. Taller, larger diameter trees will likely sustain less

Table 5-6
Relative Fire Resistance of Various Tree Species

Level of resistance	Common Name	Scientific name
Highest	Western larch	Larix occidentalis
Highest	Coast redwood	Sequoia sempervirens
Highest	Giant sequoia	Sequoia gigantea
High	Ponderosa pine	Pinus ponderosa
High	Douglas-fir	Psuedotsuga menziesii
Medium	Grand fir and white fir	Abies grandis and A. concolor
Medium	Lodgepole pine	Pinus contorta
Medium	Western white pine and sugar pine	Pinus monticola and P. lambertiana
Low	Western red cedar	Thuja plicata
Low	Western hemlock	Tsuga heterophylla
Low	Noble fir	Abies procera
Low	Engelmann spruce and Sitka spruce	Picea engelannii and P. sitchensis
Very low	Subalpine fir	Abies lasciocarpa
Very low	Pacific silver fir	Abies amabilis

Sources: Whelan 1995; Wright and Bailey 1982; and Starker 1934.

damage than smaller trees of the same species. Similarly, conifers with thicker bark and longer needles that provide shielding to large buds may suffer less damage (that is, crown scorch, crown consumption, stem mortality) than thin-barked, short-needled conifers with small buds (Miller 2000). For example, trees such as coast redwood (*Sequoia sempervirens*), Western larch (*Larix occidentalis*), Douglas-fir (*Pseudotsuga menziesii*), and ponderosa pine (*Pinus ponderosa*) develop a thick layer of dead bark with age that makes possible survival from most surface fires. Some conifers, such as coast redwood and ponderosa pine, have protected buds that permit replacement of foliage and branches consumed in a crown fire (Kimmins 1987). Tree survival also will depend on density and spacing between tree crowns, as dense, closed-canopy stands may be more likely to support active crown fires than open forests. Features of the tree crown and stand growth patterns thus also need to be considered, along with rooting habit and associated lichen growth (dangling from branches), as well as management activities (or inactivity). All of these characteristics will influence the likelihood of crowning and the severity of needle scorch or consumption. Even so, trees that survive a fire may become susceptible in later years to insect attacks, pathogens, or drought.

Plants that are killed by a fire may persist on the site if they possess adaptive mechanisms that allow the species to persist postfire. Sprouting is one such mechanism that allows plants to recover after a fire. Live shoots can originate from aboveground growing parts (stolons) or from various depths within the litter, duff, and mineral soil layers (Figure 5–13). Fire and other disturbances can stimulate sprouting in some plant species through alteration of hormone balances that control the growth of different plant parts. Also, the postfire environment may allow more light to reach the forest floor, increase soil temperatures, and thereby promote sprouting.

Fire has been shown to stimulate flowering and influence seed dispersal in some species. The reproductive success of some *Xanthorrhoea* species, an Australian shrub, is greatly enhanced by fire. Flowering may be stimulated by ethylene in smoke (ibid.). Lodgepole pine (*P. contorta*), jack pine (*P. banksiana*), and black spruce (*Picea mariana*) have serotinous cones that open only in response to heat. High temperatures melt the resinous bond that keeps the cone scales from opening in the absence of fire (ibid.).

Figure 5-13
Location of Various Plant Parts that Can Regenerate New Shoots Following Fire

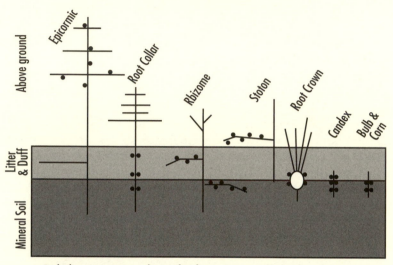

• Buds that generate new shoots after fire

Source: Miller 2000.

Nonserotinous lodgepole pine trees are thought to occur in areas that burn less frequently.

Some plant species produce seed that lie dormant in the duff or soil until a fire occurs. If these seeds are not exposed to lethal temperatures in a fire, they can germinate in response to the heat. Shrubs such as *Acacia, Arctostaphylos, Ceanothus,* and *Rhus* produce large quantities of hard-coated seeds that germinate only after heating (ibid.). Seeds close to the soil surface may be consumed or killed in a severe fire; however, these same seeds may be stimulated to germinate in a lower severity fire or if they are heated but insulated from lethal temperatures (Figure 5–14).

In reality, the plant composition following fire in an area will depend on a complicated set of factors, including the prefire species present (and their capabilities for withstanding heat), the fire's severity, and the postfire environment (for example, rainfall, solar insolation), as well as the time elapsed. Some authors (see, for example, Rowe 1983, Whelan 1995) have explained the postfire vegetation changes by classifying plants' responses according to various survival and regeneration strategies:

Invaders: Well-dispersed weedy species with short-lived seeds;

Evaders: Species with long-lived propagules stored in the soil or canopy;

Avoiders: Shade-tolerant species with slow reinvasion;

Resisters: Adults can withstand low-intensity fires, otherwise intolerant;

Endurers: Sprouting species.

Yet another strategy has been proposed by Mutch (1970), involving enhanced flammability coupled with regeneration

Figure 5-14
The Effect of Heat Exposure on Seeds

Seeds at or near the surface may be killed during flaming or by prolonged heat exposure after passage of the flaming front. Seeds lower in the duff layer may be insulated and even stimulated to germinate from the heat.

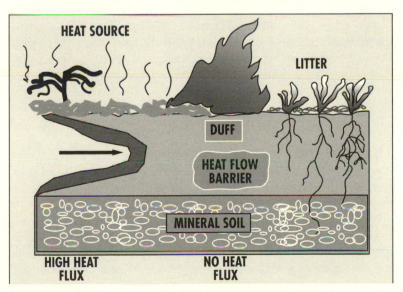

Source: Ryan 2001.

strategies (seed or sprouting). Accordingly, plant species in fire environments may burn more readily but then successfully reoccupy the postburn site, as compared with competitive species that are less fire-adapted. Some southern California chaparral species may exhibit this tendency. Others include Jack pine (*Pinus banksiana*), fireweed (*Epilobium angustifolium*), and bracken fern (*Pteridium acquilinum*), as suggested by Kimmins (1987).

Fire Adaptations and Dependence

Other fire adaptations mentioned in the literature include very rapid early growth, to elevate the terminal bud and foliage out of reach of surface fires, coupled with rapid development of thick bark (ibid.). The classic example is provided by longleaf pine (*Pinus palustris*), found in the southern United States. Its terminal bud remains close to the ground for up to five years after germination, while the seedling develops a large root system. During this time, the bud is protected from the frequent surface fires characteristic of the region by long, fire-resistant needles that form a dense circle around the bud (the so-called grass stage of the tree). The grass stage ends when the rooting system has developed sufficiently to support rapid height growth by the seedling, carrying the fire-sensitive bud above the reach of surface fires.

Kimmins (ibid.) adds that longleaf pine is not only resistant but also dependent on fire. The tree is susceptible to a foliage fungus that is eliminated or retarded by fire, so managers may apply prescribed fires during the grass stage to control the pathogen. Furthermore, fire will eliminate or reduce competition from less fire resistant understory hardwood species that would invade the site if fire were excluded.

Fire Severity

Fire severity is an important concept for understanding heat effects on organisms and natural resources, by categorizing the degree to which fire alters vegetation and ecosystems. Although no specific measurement units are implied, the concept is still extremely useful. We use the term "fire severity" in several contexts. When speaking of fire regimes, we may use the term to describe the characteristics of historical fires that burned a particular vegetation type, plant community, or biome in an area or geographic region. We also use the term "fire severity" to describe the ecological impacts of a fire or set of fires in an area—for example, with reference to flora and fauna, to soils and water, or the atmosphere and archeological resources.

In assessing fire effects, fire severity is more important than fire intensity, since the latter is a measure of the rate of heat release in the fire's flaming zone—and can be used in calculations for flame length (that is, equation 9). It may not, however, account adequately for heat transmitted after passage of the flaming front or downward into the soil profile. By contrast, "fire severity" encompasses the ecological effects of both fire intensity (aboveground) and depth of burn (belowground). Severity thus better incorporates both short- and long-term aggregate effects of flaming and glowing combustion—for example, differentiating the effects of short flame lengths followed by lengthy glowing combustion versus taller flames followed by shorter glowing combustion. These distinctions are important, since fire effects will be vastly different in ecosystems experiencing contrasting time-temperature exposures at different heights above and below ground. For example, Ryan (2001) demonstrates that fire's ecological effects will be quite distinct because of time-temperature profiles above- and belowground in (a) savanna or grassland fire (with no duff consumption); (b) crown fire with no duff consumption; (c) ground fire with duff consumption; and (d) crown fire with duff consumption. Dramatically different fire effects will be expected under these contrasting conditions. Sensitive growing tissues and reproductive parts (buds and seeds) will react to these differential heat pulses accordingly. Thus, savanna and crown fires (with no duff consumption) might kill or damage aboveground vegetation, yet provide an advantage to plants that reproduce vegetatively by sprouting in juvenile or adult stages. Plants that reproduce by seeds stored in the ground might also gain a foothold in the postfire stand. By contrast, ground fires with little flaming and crown fires (that is, that burn the duff layer) may consume subterranean seeds yet provide an opportunity for plants that reproduce by seeds dispersed from outside the burn area to occupy the postfire site. The ground fire may also favor plants that store seed in crowns. Again, fire intensity describes only the heat pulse aboveground that contributes to active flaming, so at best it permits incomplete inferences about heat effects.

Postfire evaluation schemes attempt to systematize the collection of quantitative and visual cues for measuring fire severity, for possible linkage to observed fire effects. For example, Ryan and Noste's (1985) methodology involves assignment of burn conditions within a two-dimensional matrix for classifying the heat pulse received by aboveground plants and the heat pulse transmitted down into the soil. With modification, their approach also has proven useful for the evaluation of fuel treatment effects on wildfire severity (Omi and Kalabokidis 1991, Pollet and Omi 2002, and Omi and Martinson 2002).

Perception of fire as either harmful or beneficial depends on your personal values, interests, or point of view. For example, a timber manager, wildlife habitat specialist, hydrologist, and recreationist may have drastically different opinions about the effects from any single fire, each focusing on his or her own particular point of interest. Thus a wildlife specialist may be pleased with increases in postfire food sources for small mammals, while perhaps not being so concerned about reductions in the density of trees in the area. Or a hiker may be overjoyed about the improved hiking when vegetation is cleared along stream crossings, perhaps overlooking the increases in stream temperatures and fish dieback accompanying the removal of trees along the creek banks. Fire ecologists and managers thus must be aware of prevalent and conflicting points of view as the forest recovers from a fire.

Out of long tradition, wildfires are commonly considered "bad" and prescribed fire as "good." Actually, some fires are declared "wild" that may in fact benefit certain ecosystems in terms of nutrient recycling, wildlife habitat improvement, or other changes that occur within the burn perimeter. Conversely, prescribed fire effects depend on the quality of execution. Appropriate execution will likely mimic the more benign impacts of wild fires; poor execution of prescribed fires results in damage. Fire management and the public will have come a long way when they can accept the beneficial aspects of wildfires and reject the damaging results from "bad" prescribed fires. Generally, intentionally set prescribed fires are preferable to random wildfires, because managers exert control over the timing and location of burning to meet prespecified objectives. Nevertheless, fire severity is the appropriate criterion for judging the ecological effects of fires, regardless of whether human-caused or the result of natural causes.

The discussion of fire effects is a huge field that goes far beyond the scope of the current discussion. Interested readers may wish to pursue numerous literature sources available in local libraries or on the worldwide web. Fire effects on vegetation are summarized in Brown and Smith (2000), including effects on individual plants and fire regimes. Smith (2000) summarizes effects on fauna, including wildlife and nongame species; Sandberg et al. (2002) summarize the effects on air resources resulting from burning. Additional summaries are noted in Chapter 7, along with valuable sources available on the Internet, such as the Fire Effects Information System (listed under Nonprint Resources) and the predictive First Order Fire Effects Model (see Software).

Fuel Consumption

The rate and degree of fuel consumption will determine the effects of a fire whether wild or prescribed. Fuel consumption affects the soil substratum, the microenvironment, and the amount of plant sprouting or seed regeneration that occurs in an area. If consumption can be managed, so can a fire's effects. On a wildfire, consumption is unmanaged until firefighters or weather conditions remove heat, fuel, or oxygen from burning. Wildfires are considered so damaging precisely because fuel consumption is not managed (including houses in the area) but left to chance and circumstance.

In reality, a fire only consumes a small portion of aboveground biomass. For example, as little as 5 to 10 percent of stems and branches may be consumed in a high-intensity fire in a Douglas-fir forest in the Pacific Northwest, while nearly 75 to 100 percent of foliage, forest floor, and understory may be consumed (Agee 1993). Thirty percent of snags and downed logs may be consumed, depending on the degree of rot, although all of these figures may fluctuate considerably during drought or wet periods, in different vegetation types, or in other geographic regions.

Prescribed fire consumption can be managed in several ways, including specification of allowable wind and fuel moisture conditions under which ignition will occur (that is, in the prescription). Higher fuel moistures, in particular, will retard consumption by absorbing heat and lowering fire intensity. Other options for regulating prescribed fire consumption include selection of the appropriate ignition technique and pattern, and restriction of the smoldering combustion stage. Possible ignition techniques vary from the handheld drip torch, to truck-mounted terra-torches, to aerial devices (for example, helicopter drip torches and Ping-Pong balls filled with potassium permanganate). Fires can be ignited in strips of varying width, and with or against the wind and slope—all of which will alter the fire severity and consumption. Smoldering consumption can be abbreviated by mopping up smoldering fuels or by burning when the large diameter fuels and duff have higher moistures—for example, in the spring.

The greatest degree of fuel consumption occurs during the smoldering combustion stages of a fire, after passage of the flaming front, when heat is transferred mostly by conduction. Guidelines for regulating fuel consumption can be obtained by using the CONSUME program (Ottmar et al. 2002). For example, in the Alaskan boreal forest, forest floor consumption can be predicted from the following equation:

$$y = 0.27501 - 0.04601 \, x_1 + 0.08258 \, x_2 \tag{10}$$

where,
y = forest floor reduction (cm);
x_1 = upper forest floor moisture (%); and
x_2 = preburn forest floor depth (cm)

Equation 10 was developed from eleven experimental burn units, explaining 82 percent of the observed variation in forest floor reduction (ibid.). Similar relationships are embodied for different fuel types and geographic regions in the CONSUME software.

A better understanding of fuel consumption principles allows managers to fine tune and calibrate fire prescriptions for an area. As experience is gained in understanding how fuel consumption can be controlled, the manager becomes more skilled in using fire to produce the desired changes in the ecosystems being treated while minimizing risk.

Emissions and Air Quality

Fire emissions (particulates, gases, and other products of combustion) are among the most critical of fire effects to be managed, especially as smoke is associated with greenhouse gas formation and global warming. Smoke also contributes to regional haze and may present human health hazards to individuals with pulmonary disorders (such as asthma) or allergies. Globally, as much as a third of the total emissions from biomass burning may be associated with burning in savannas (grasslands with a scattered cover of trees and shrubs) in Africa, South America, and Australia. Haze associated with the 1997 fires in Indonesia, Malaysia, and Singapore caused billions of dollars' worth of commercial disruptions. In the United States, the federal Clean Air Act mandates protection of air quality, especially around class I (that is, pristine) areas, though enforcement is left to individual states. State jurisdiction for enforcing clean air standards extends to federal lands such as national forests and parks.

Key fire contaminants include particulate matter (PM 10 and PM 2.5 refers to particle size, in microns (μ) or 10^{-6} meter), elemental and organic carbon, volatile organic compounds, sulfur oxides, nitrogen oxides, nitrates, and carbon monoxide. These contaminants vary in terms of visibility impairment, contributions to regional haze, impact on human health, longevity in the atmosphere, and nuisance effects—though particulates generate the greatest concern from a fire standpoint. Individual states regulate air quality standards, while the Environmental Protection Agency has developed national ambient air quality standards (NAAQS) to regulate effects on human health.

Particulates suspended in the air produce the primary impacts on air quality (aesthetics, health, and safety). Aesthetic impacts include plume blight and haze resulting from light scattering by particulates. Recall that the greatest proportion of particulate diameters produced in wood smoke (from Figure 5–9) are in the smallest size classes, less than 1 micron (μ) in diameter. In fact, as much as 80 percent of wood smoke particles are between 0.3 μ and 0.7 μ, making them very efficient "scatterers" of visible light (Deeming 1988). These particles contribute to unsightly smoke in the air when the fire is burning, and they may remain aloft for an average of ten days, causing haziness in the sky. Essentially all particles of less than 10 microns can be respired into the human body. Larger particles may be collected by hair follicles and mucus in nasal passages, and cause coughing and discomfort. Smaller particles, of less than 1 μ (82 percent of the particles in wood smoke, Figure 5–9), may go deeper into the lungs and become entrapped in the alveoli (air sacs), causing bronchial discomfort, illness, or disease. Human health impacts depend on a variety of risk factors, including age, general state of health, and the toxicity of the gases and particles in wood smoke. Adults with chronic respiratory disease or asthma and children with bronchitis, asthma, or hay fever are especially susceptible, although the general public also may suffer. Firefighters themselves are susceptible to risks associated with prolonged exposure to smoke while working on the fireline.

One particular risk to firefighters is carbon monoxide (CO) poisoning. CO is produced in greatest quantities during the smoldering stage of a fire, when combustion is least efficient. CO attaches to hemoglobin in the bloodstream and displaces necessary oxygen flow to the brain, causing dizziness, disorientation, or, in the worst cases, suffocation. Even though the likelihood of CO poisoning dissipates quickly with distance from the source and as oxygen is replenished, at the end of a long shift with physically arduous activity, any drop-off in mental acuity can be harmful to a firefighter—especially when quick decisions are necessary.

Managers attempt to mitigate air quality risks by controlling smoke production, transport, and dispersion. Excessive smoke from wild and prescribed fires can threaten public safety by reducing visibility along highways or air transport routes. Serious chain-reaction automobile collisions involving hundreds of vehicles have been documented when unwary drivers have attempted to travel through smoke-filled highway corridors. Posted warning signs may help, but drivers need to exercise common sense when smoke obscures the road ahead.

Postfire Effects on Soils/Hydrology

The effects of wild and prescribed fire on water runoff and soil erosion vary greatly, depending in part on fire severity, but also because both the vegetation canopy and the physical properties of the soil can be affected, as well as protective duff and litter layers atop the soil (MacDonald and Stednick 2003). Wildfires can increase the amount of water runoff and soil erosion by several orders of magnitude (Wagenbrenner 2003). Erosion rates from undisturbed forests are generally quite low, with estimates for the western United States ranging from 20 to 1,200 kg/ha/yr. Wildfires increase the amount of runoff and erosion, because of vegetation cover disturbance and removal. The risks of heightened erosion can last anywhere from one to five years following the fire, depending in large part on the intensity and location of rainstorms in relation to the burned area. In midelevation forests of the Colorado Front Range, overland flow and surface erosion are initiated after wildfires whenever rainfall intensities exceed 10 mm/hr (MacDonald 2004). The process is quite complex and variable over an area, but it starts with plant death and damage resulting from high-severity fires. As plants die, their stems and root systems no longer are available to stabilize soil movements.

The resultant reduction in plant cover will reduce canopy moisture interception. Removal of plant and litter cover increases soil evaporation rates, and allows more rainfall to land directly on bare soil (Wagenbrenner 2003). This increases soil sealing and reduces infiltration (DeBano et al. 1998), making soils more erosive and increasing overland flows. In coniferous and chaparral forests, high-severity fires may cause a hydrophobic (or water-repellent) layer to develop below the soil surface, further reducing infiltration and leading to soil layer movement (DeBano 1981). This water-repellent layer in the soil lasts one to two years following both wild and prescribed fires, with erosion rates five to twenty-five times greater following high-severity burning in comparison to sites burned at moderate or low severity (MacDonald 2004). If saturated by rainfall, the soil above the hydrophobic layer can move or slip downhill on account of gravity. The surface soils on whole hillsides may literally liquefy and flow downhill, threatening communities and downstream water storage facilities. Such mass movement may even occur on dry slopes in steep terrain following a wildfire.

Thus wildfires followed by intense rainstorms can be extremely damaging, resulting in massive flooding and overland flows of sediment, as well as scouring of creek channels and deposition of soil and debris in rivers and reservoirs. Large, high-severity wildfires pose a much greater threat to water quality than do low-severity wild or prescribed fires (MacDonald and Stednick 2003). Such damage is especially acute if this so-called fire-flood sequence occurs in the vicinity of homes or structures, or if sediments flow into reservoirs, reducing storage capacity and the quality of drinking water. Such episodes have accentuated the short- and long-term effects of severe burning, especially in the southwestern United States. For example, intense Christmas day rainstorms following the southern California fires of 2003 contributed to massive debris flows (mud, boulders, trees) into Waterman and Lytle Creek canyons in the San Bernardino National Forest. Sadly, many fatalities resulted when mudslides buried unsuspecting children and adults camping downslope from a recently burned area. This disaster might have been predicted, based on historical records indicating dozens of previous wildfire-rainstorm-debris flows in southern California. Perhaps the most infamous example occurred just after New Year's Day in 1934, when a wall of mud and rocks 6 m tall swept through the towns of La Crescenta and Montrose, killing forty-nine people (Bustillo 2003).

Some atmospheric scientists have suggested that large burned areas may actually facilitate the development of intense convective storms, thus triggering or contributing to increased runoff and erosion. A possible explanation for this phenomenon is that burned soils have lower albedo and absorb more heat, thereby raising aboveground temperatures because of long-wave radiation and convective heating. Although not systematically studied to date, increases in summer storm runoff have been documented following the 1994 South Canyon fire, the 1996 Buffalo Creek fire, and the 2000 Bobcat fire, all in Colorado. This effect may not be apparent in higher elevations or during droughts (MacDonald and Stednick 2003), but it deserves further study.

Fire Protection and Management

All fires, from the smallest to the very large, go through roughly the same basic phases, from initiation through growth and then extinction. During the initiation phase, an idealized fire grows in size as the fire spreads away from its point of ignition in response to fuel, weather, and topographic variations. During the growth phase, the fire accelerates relatively rapidly in size but eventually reaches a point where growth increases, but at a decreasing rate. Eventually growth slows and the fire size levels off and finalizes. Individual sectors or pockets within the fire perimeter may continue to burn beyond this point, so firefighters need to mop up actively burning portions to prevent the fire from escaping control. Although this pattern generally holds for all fires, eventual size and growth rates vary considerably. Some fires may go through several acceleration-deceleration growth phases. Blowups and major fire runs accelerate fire growth, management activities may slow the fire's growth, or natural influences (such as winds and fuel moisture) may increase or decrease growth rates.

Most fires are fought with suppression crews and equipment with the intent of keeping fires as small as possible, but the same trends prevail in terms of initiation, growth, and extinction. Even intentionally ignited prescribed fires or fires that are allowed to burn to meet management objectives will exhibit these same general trends in fire growth and eventual size. When wildland fires are used for resource benefit, the final size may be predetermined by the maximum management area (MMA) set in advance by planners. In fact, prescribed fires or those allowed to burn for resource objectives may be encouraged to treat as large an area as practical or as environmentally acceptable.

A fire that ignites in the forest must first be detected, then located geographically and communicated to the local fire dispatcher. Depending on accessibility, available initial attack forces are dispatched to the incident and begin suppression efforts, such as digging a fireline or knocking flames down with water hoses. Thus managers have several options for keeping fires small when objectives call for aggressive fire control (Figure 5–15). For example, managers can focus on minimizing the time until the fire is detected (t_D), reducing the time until initial attack forces arrive at the fire (t_I), or restricting the overall time required to extinguish the fire (t_E). Obviously, fire size can be reduced (along with associated expenditures) if t_D and t_I are kept small. Detection times can be reduced by investments in fire detection equipment, aircraft, or in educating the public about reporting ignitions to responsible authorities. Reductions in time until initial attack may be possible if more crews and suppression resources (for example, aerial retardant planes) are hired and trained to respond to

Figure 5-15
Fire Growth over Time
Over time, fire growth exhibits a similar pattern for all fires, including initiation, growth, then extinction. Options for reducing fire size include reductions in detection time (t_D), initial attack time (t_I), and overall time until extinction (t_E).

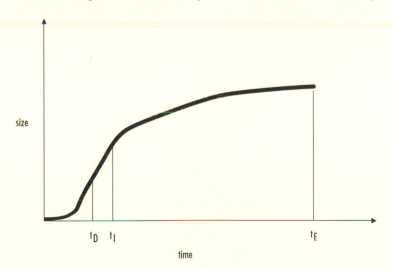

fire dispatches. Investing more in suppression resources can reduce extinction times (t_E), which also may be facilitated by fuel treatment expenditures that reduce the likelihood that fires will require costly extended attack—that is, after a failed initial attack. Alternatively, investments in fire prevention efforts aim to reduce the likelihood of damaging fire ignitions.

The prevention of large, catastrophic fires deserves special attention, since those incidents involve unusually high costs and cause massive damage to natural and human environments. The mechanisms associated with large fires are conceptually straightforward (Deeming 1990): flammable materials (fuel) are exposed to an intense heat source (firebrand) coincident with conducive weather and fuel conditions (fire danger). Reduction of large fire probability thus focuses on one or both of two strategies: (1) elimination or reduction of firebrand sources (risk management); or (2) removal or modification of the fuel to reduce its flammability during high or extreme fire danger conditions (hazard management).

Risk and hazard are important terms to distinguish: Risk is a wildfire causative agent, such as lightning, arson, or railroad trains; it also refers to the likelihood or probability of ignition in an area, again by a causative agent. Hazard describes the properties of a fuels complex related to ignition susceptibility, wildfire behavior and severity, or suppression difficulty (Omi 1997). These definitions are not universally acknowledged or used uniformly by all in the fire management community, although most agree that risk and hazard need to be distinguished. For example, in some engineering applications risk is defined as the product of an event probability times the value at risk.

Fire managers have long realized that risks cannot be eliminated totally, but at best they should be acknowledged and managed commensurate with fuel hazard levels and values-at-risk. Thus some level of risk must be accepted under most circumstances. After all, forests have burned for millennia, with no end in sight. Where values are high and risk cannot be reduced sufficiently, the management focus on fuel hazard reduction may have greater payoffs than all-out suppression.

In addition, community vulnerability or ecosystem susceptibility to fire may be of concern to fire managers. Developed communities may be vulnerable because of their location nearby to fire-prone environments. Ecosystems may be susceptible when fuels have accumulated because of fire exclusionary practices or as a result of vulnerability to other types of disturbance (Chapters 1 and 2).

Systematic Fire Management

Back in the early twentieth century, the tasks confronting forest managers were straightforward, even if sometimes overwhelming. Plain and simple, fire control was the first responsibility of every forest ranger. With the passage of time, and as forest management became increasing complex, fire control became its own specialty, although the major focus remained unchanged. Fast-forward to the present, and today's fire manager must focus not only on controlling fire but also on using fire as a tool, with the goal of ensuring ecosystem sustainability—almost as if the arch-enemy has now become an essential ally. However, the balance between fire control and fire use is not easily struck, as there is still far too much unwanted fire in comparison to useful fire. And the distinctions sometimes blur, as for example during the infamous Cerro Grande prescribed fire escape near Los Alamos, New Mexico, in 2000 (Chapter 1). Moreover, the responsibilities of fire managers have grown more complex, with the need for greater attention to environmental concerns and accountability, among other demands placed on the job. So the modern fire manager requires a completely different set of skills and competencies than did the early forester or fire control officer. Some of these skills may be apparent from the following list of activities that are part of typical systematic fire management efforts for an area. However, as noted below, the required competencies go much beyond the activities listed.

Fire prevention: Reduce unwanted ignitions, including firebrand production (agents) and lower susceptibility of fuelbeds to ignition (that is, risk).

Urban interface management: Manage communities, dwellings, and structures near wildlands.

Fire detection: Detect a fire when small and report to appropriate authorities.

Fire preparedness: Advanced preparation in anticipation of future fire outbreaks.

Dispatch: Process as quickly as possible reports of fire location, available forces, and sending of resources plus support to a fire.

Fuels management: Reduce fuel hazards consistent with land management objectives.

Wildland fire use: Employ or herd fire to achieve land management objectives.

Fire suppression: Initial attack, extended attack, and mop-up to extinguish a fire.

Communications: Make sure information and instructions are communicated and understood, using best available technologies.

Safety: Protect firefighter lives and the public, as well as natural resources.

Fire rehabilitation: Restore plant cover, protect soil and water resources.

Ecosystem restoration: Restore ecosystem structure and function.

Time and space do not permit detailed discussion of each of the activities in the above list. Highlights are summarized below, however, mostly to provide a flavor for the scope of concerns and interests confronting contemporary fire managers.

Fire Prevention

An old fire-control adage proclaims that a fire prevented does not need to be suppressed. By the same token, costly fire suppression activities may be avoided. Smokey Bear became the poster child of national fire prevention campaigns when, as a cub, he was plucked from the ashes of a New Mexico wildfire in 1950. For decades, his message, "Only YOU can prevent forest fires!" provided a motivation for keeping fire out of the forest. Of late, Smokey's message has had to be altered in light of scientific evidence about the harmful effects of complete fire exclusion in some forest types—for example, long-needled pine ecosystems that historically experienced frequent, low-intensity surface fires. In such forests, prescribed forest fires may represent a long-term alternative to damaging catastrophic fires.

The more traditional view toward fire prevention relies on the so-called three Es of Education, Enforcement, and Engineering as cornerstones of fire prevention efforts. Sample education efforts include public information spots in the news media, interactive websites (see, for example, http://www.fs.fed.us/fire/links/links_education.html), or presentations in grade schools. Enforcement efforts attempt to curb risky behavior by implementing laws or regulations. For example, forests may be closed, or heavy equipment may be shut down during periods of high fire danger. Industrial logging operations may be prohibited early in the afternoon on high fire danger days during the summer. Law enforcement agents may increase forest patrols in high-risk areas on days when Red Flag warnings are issued by fire weather forecasters. Engineering solutions include remote surveillance of lightning or human fire starts, and efforts that remove ignitable biomass. Alternatively, fire-safe environments may be created, such as thinned forest buffers or defensible spaces around homes and communities (Dennis 1992).

A more progressive view toward fire prevention examines the psychological and sociocultural motivations for fire-setting and may attempt to alter human behavior, using sociological techniques (Doolittle and Lightsey 1980). For example, in the South, woods-burning stems from long-standing traditions and beliefs about firing the forest, whether for clearing undergrowth, eliminating pests, or facilitating wildlife, cattle, or timber management. Community values clearly support burning, so fire prevention campaigns need to be directed toward fire deviants without alienating local opinion. In such settings, rules and laws may be unenforceable and less effective than working with local opinion leaders.

Modern practitioners also recognize that all ignitions may not need to be excluded or prevented from some ecosystems; in fact, complete prevention of fires may lead to undesirable consequences, such as increased fuel loadings and undesired species composition. Some fires might be beneficial under prescribed circumstances. This view presents several problems to managers, in terms of screening to identify those ignitions that might be more beneficial if allowed to burn. An even more intractable assessment involves determining the extent to which fire prevention activities, such as increased law enforcement efforts, actually reduce human ignition sources.

Another aspect of fire prevention includes fire forensics, or systematic investigation into the point of origin and ignition cause. Determination of fire origin and growth may be important for assigning ignition responsibility and for settling damage claims. For that reason, firefighters are instructed to protect or isolate a fire's point of origin whenever possible, whether a lightning struck tree, a cigarette butt, or other cause. Guidelines for carrying out a fire forensic investigation are noted in DeHaan (2002), though the procedures for wildland fire are quite limited. Individuals interested in fire cause investigation require specialized training in recognizing fire growth patterns. For example, fire char will typically be higher on the lee side of a tree trunk in the path of a spreading fire, much as fire scar formation is more likely on the uphill side of the tree bole. Also, a fire ignited at the bottom of a hill will typically spread uphill with an inverted U- shape. These and other patterns, such as needles frozen in the direction of fire spread, can sometimes be used to track a fire's growth backward to the point of origin.

Urban Interface Management

Fire prevention in the so-called urban-wildland interface is best implemented in collaborative partnership among residents, landowners, communities, county regulators, and fire agencies, among others. Many wildland areas in the West, where fire is both natural and common, have become interspersed with single homes, residential areas, or entire municipalities. Often communities are developed and homes are built without regard for wildfire risks and hazards. In the absence of concerted partnership efforts, insurance companies might boost premiums, reduce coverage, cancel existing policies, or drop fire insurance altogether—as reportedly occurred in certain areas of California following the 1991 Tunnel fire in the Oakland-Berkeley Hills.

Homeowners can protect themselves by clearing vegetation around the house (creation of a so-called defensible space), using fire resistant construction materials for roofs, siding, and decks, and landscaping with greenbelts and drought-resistant native plants. Communities and county regulators can adopt sensitive and sensible subdivision design criteria. Fire agencies can work with all concerned in developing community-based fire education and inspection programs.

Creation of defensible space can reduce the risk of wildfire while providing firefighters with extra margins of safety. Some of the more common features of defensible space include the following:

Noncombustible roofing materials;

Creation of fuelbreaks (low-volume vegetation) and firebreaks (vegetation-free zones);

Thinning or elimination of flammable plants around structures (for example, pines, eucalypts);

Removal of woody fuel, leaf, and pine needle jackpots;

Removal of trees near to structures;

Pruning of branches or raising crown base heights that overhang roofs and decks;

Reduction of fuel loads in zones surrounding the home;

Installation of irrigation systems or water storage facilities;

Maintenance of good access roads and turnaround areas.

Additional guidelines for creating fire-safe communities can be accessed at the FIREWISE website (Chapter 7) or by entering "living with fire" into any Internet search engine.

The urban interface fire management problem is not restricted to the United States. Housing losses occur as a result of wildland fires, but they are not so numerous in Japan, Australia, and Canada. Losses are less extensive in southern Europe in part because of greater reliance on nonflammable building materials for home construction. Still, fire losses do occur. In Australia, homeowners are encouraged to "prepare, stay, and survive" as an alternative to the massive fire evacuations (that is, involving tens of thousands of people) ordered annually in the western United States.

Displaced homeowners commonly express frustration with evacuation procedures that are implemented unevenly or executed poorly. This frustration can be ameliorated partially if homeowners form partnerships with public officials to understand fire risks, participate in hazard mitigation activities, and plan evacuation procedures before a fire occurs.

Detection

A fire lookout tower rising above a prominent ridge or vantage point provided an early fixture on the landscape as formerly wild forests came under management in the United States during the twentieth century. These towers also stood as symbolic sentinels standing watch over the vast woodlands of the United States, with public servants ever on the lookout against random lightning or human ignitions. These threats became especially pronounced as forest resources were increasingly exploited, aided by railroads extending general access to frontier communities while facilitating the extraction and transport of raw materials (for example, lumber and ores) at the end of the nineteenth century.

As development of the frontier proceeded, steam locomotives would often eject burning cinders and sparks along the railroad right of ways, starting fires that extended into the forest where loggers had previously left debris (treetops, limbs, and other slash) from cutting operations. Thus railroad ignitions under hot, dry conditions would transform the woods into blazing infernos. In particular, the 1903 and 1908 fire seasons in the Northeast and the West spurred public outcries for the creation of forest fire control systems.

The first fixed detection lookouts were primitive, being staffed by observers perched atop makeshift log decks from some open vantage point. Observers would remain vigilant during daylight hours, watching for ignitions during the fire season, especially in "hotspots" such as lightning alleys or travel corridors. Over the next several decades, those crude detection stations evolved into the more familiar metal towers with enclosed cabins rising above the forest floor, equipped with fire finders (crosshaired sighting windows) that allowed the lookout observer to fix the ignition azimuth and legal description. A sighting from a second lookout with a different vantage point allowed triangulation of the fire's location precisely. Eventually, telephones or heliographs (mirrors for signaling Morse code) gave way to shortwave radios for communicating sightings to the dispatcher (Fuller 1991). In addition to fire detection, some lookouts in coastal forests were trained to provide early warning for anticipated enemy air attacks during World War II.

Today, aircraft flying systematic patrol routes over high-risk locations have become standard in some remote locations, more recently equipped with sophisticated heat-sensing detectors and global positioning technologies. Satellites equipped with radiometers to fix locations and send out automatic alerts to fire dispatch centers are used in some countries, such as Finland, Sweden, Norway, Estonia, Latvia, and Russia (Kelha et al. n.d.). In less remote areas, rotating video and digital cameras with wired or wireless connections to a central observation center may be used instead of a human observer.

The aim of lookouts and other detection systems (such as aircraft) has always been to detect fires as soon as possible while fire size is still small, so that suppression forces can be dispatched to an incident as soon as possible. Advancing technologies have spawned numerous innovations. For example, fixed-location sensors that recognize the CO_2 signature from an ignition can be linked to a local dispatch center or else can trigger deployment of heat shield curtains to protect nearby houses. Mobile sensors aboard detection aircraft have become important for mapping fire perimeters and identifying hotspots burning outside of control lines. Aircraft also may be used to patrol areas in search of fires after a rash of cloud-to-ground lightning strikes. Aircraft patrol routes and flight lines are planned to prioritize coverage of lightning strike concentrations, perhaps detected by automated lightning detector sensors. Sometimes these ignitions might not be apparent for a day or so, until heat, flame, or smoke becomes detectable. These delayed fires are sometimes referred to as sleeper fires, also known as holdover fires.

Satellites and aerial photos also provide useful information for mapping fire perimeters and subsequent damage. After a fire has burned over an area, satellite sensors and aerial photography may be used for a variety of purposes. Conventional Landsat thematic mapper and multispectral imagery have proven useful for a variety of applications, such as mapping vegetation mortality, soil moisture, burn severity, and forest canopy damage (Williams 2003). Severely burned areas may be candidates for burned area rehabilitation measures, such as seeding or mulching to reduce soil erosion. Remotely sensed data also can be used for assessing forest regeneration after a fire, or to detect changes in wildlife habitat.

Satellite imagery generally makes use of different spectral bands or electromagnetic reflectance, such as near- or mid-infrared wavelengths that penetrate smoke but not forest canopies equally readily. Laser altimetry, or lidar (light detection and ranging) technologies, shows promise for remotely sensing forest structures beneath tree canopies, but its fire detection capabilities are unproven, especially over large areas.

In areas with high population densities, the lookout tower has been decommissioned or replaced because of reduced air quality, which hinders observations. Some towers have been shut down because of the high upkeep cost, or the evolution of more efficient alternatives for detecting and reporting fires, such as satellites or hikers and motorists with cell phones. In essence, more cost-efficient devices have replaced the human observer, often viewed as the weak link in the system. Today many lookout towers have been refurbished to become important destinations open to the public for weekend sleepovers, day hikes, or recreational drives.

Dispatch

Dispatch facilities are like the nerve centers of the fire organization, processing information from several sources and then making sure that appropriate responses are undertaken. Forest fire dispatch centers process reports on an ignition by lookouts or other detection sources, then facilitate the movement of available forces. Dispatchers also communicate with respondents to a fire, keeping track of time and resource transport as well as responding to requests for additional support. These requests may be filled from other local resources, or the dispatcher may need to pass the request up the organizational hierarchy—perhaps to a geographic area command center. Requests that exceed available resources at this level may be forwarded to the National Interagency Fire Center. In all cases, the dispatch office will be at the center of the two-way communications between the field and higher-ups. When multiple ignitions occur simultaneously (for example, because of lightning or arsonist outbreaks), or when complex fires involve multiple jurisdictions, the dispatch office may become a hub for control of ground and air traffic.

Depending on the organization, the dispatch may be automated or rely on automatic responses. For example, preplanned dispatch responses (of crews, engines, or air tankers) may be keyed to specific thresholds of fire danger. During peak periods of fire danger, resources such as crews or air tankers may be prepositioned or placed on standby status as a precaution against fire outbreaks. These various schemes may be developed and rehearsed in the off season for implementation during the peak season.

Dispatchers also may be responsible for communicating fire alerts or Red Flag warnings to field crews, using standard or emergency radio prompts. The dispatch office also may communicate with field crews about fire bans or closures affecting various user groups.

Preparedness

Fire managers may spend considerable time during the off season becoming better prepared to contend with upcoming incidents. Activities include training, equipping and pre-positioning of firefighting resources, equipment maintenance and replacement, fire planning (including budgeting), or writing fire management plans. Cooperative agreements may need to be written or negotiated with interagency or private cooperators. Fire specialists also may be called upon to interact with specialists (for example, soils, water, wildlife, timber, recreation, rangeland ecologists) in developing environmental assessments and forest plans.

An extensive slate of fire training opportunities are required and offered to develop basic competencies and upgrade skills within the fire workforce. Training opportunities run the gamut from basic fire school for incoming seasonal firefighters offered at the local level to advanced incident management techniques offered only to seasoned veterans at a national academy (see Chapter 7). Agency employees keep fire task books signed off by supervisors that identify training and experience, which will allow advancement within the fire management hierarchy.

Each fire station maintains a cache of firefighting tools and supplies, for outfitting engines and crews with necessary equipment. During the off season, tools and supplies are repaired, replaced, or refurbished. Preattack plans may be developed, laying out the location of important fuelbreaks and firebreaks, access roads, and water storage facilities within the area of protection jurisdiction. Plans may be developed for pre-positioning of suppression resources during periods of high fire danger, or for complying with protocols for requesting support from adjoining districts or regions.

Fire planning takes place on several levels—for example, building and justifying an annual budget request or making sure that specific projects (for example, fuels treatment or prescribed burn) are carried out as effectively as possible at least cost. The standards used in planning will vary. For example, the minimization of cost plus net value change (damages net of benefits) may be the standard for fire planning. Individual projects may be subjected to interdisciplinary review and effectiveness compared (relative to costs) with alternative treatments. Extensive use of computerized fire models (Chapter 7) and geographic information systems may be required to develop fire plans, based on costs and effects of simulated alternatives. On another level, monetary resources (such as fire severity funds) may become available on short notice, requiring submission of written plans for spending the money to beef up forces for an unanticipated early fire season. In any case, the best prepared managers are those who have invested the time to anticipate or plan for upcoming activities.

Fuels Management

Fuels management consists of activities undertaken prior to the onset of a fire to ameliorate subsequent fire effects so as to achieve land management or fire protection objectives. Intentionally or not, fuels are manipulated as a consequence of many management activities, including timber harvests, wildlife habitat improvement, livestock grazing allotments, or ecosystem restorations. By far, the best known and practiced applications are for fuel hazard and risk reduction, which may involve the following activities (after Brown and Davis 1973):

Remove all ignitable fuels within small, high-risk areas to eliminate ignitions from known sources.

Remove all fuel in a strip surrounding high-risk areas to isolate and limit the spread of fires starting inside those areas.

Reduce all fuel in a strip surrounding high-value or high-hazard areas to exclude fires starting outside those areas.

Reduce or remove fuels in strategically located blocks to augment natural fire barriers.

Eliminate fine and intermediate fuels—less than 1 inch (2.54 cm) diameter—from extensive areas to reduce wildfire spread rates.

Eliminate fuels extending from the ground to the tree crowns (ladder fuels), to reduce the danger of crown fires.

Remove standing dead trees (snags) that can produce embers (firebrands) that can be readily carried across fire control lines.

All of the above measures may be applied within or around individual houses or communities to reduce fuel hazards. Technically, fire behavior and effects can be changed by alterations to any or all of the various fuel strata, including the duff and litter; dead and downed woody, shrub, and low vegetation; and ladder and forest canopy layers. Treatment alternatives include hand piling, tractor piling, mechanical crushing, mastication, burning, chemical desiccation, and biological reduction using livestock animals, to name just a few. Different techniques fall into the following general categories, with listed advantages and disadvantages (from Omi 1997):

Disposal on site (for example, prescribed burning)

Advantage: Mimics natural processes such as fire and decomposition

Disadvantage: Air quality concerns, risk of escape, nonuniform fuel reduction

Redistribution on-site (for example, cut and scatter tree limbs and branches)

Advantage: Reduces fuel jackpots, lowers fuel depth, and facilitates natural decomposition

Disadvantage: Resulting fuelbed may be more continuous

Physical removal (for example, using crews or specialized machinery)

Advantage: Removed material may be used (for example, firewood, energy cogeneration)

Disadvantage: Possible site disturbance, nutrient depletion, cost

Type conversion (for example, changing plant cover)

Advantage: Reduces flammability by changing fuel type

Disadvantage: Native species may be displaced, use of chemical or biological methods may be questionable

Isolation (for example, by constructing fuelbreaks or firebreaks)

Advantage: Protect high-value areas or provide anchor points

Disadvantage: Landscape aesthetics may be compromised

In addition, different land management practices and policies will change fuel profiles in an area. Thus practices such as livestock grazing, timber management, water/soil stabilization, preservation, and management for ecosystem sustainability all can imply different levels of fuel hazard. In fact, the policy of fire exclusion may have had the greatest impact on fuel hazards in Western forests during the last century.

Fuel treatment traditionally has focused on wildfire hazard abatement, either by physically reducing, rearranging, or removing flammable biomass from the fuels complex. Primary contemporary methods include prescribed fire, mechanical thinning, or both. Other alternatives include biological (for example, grazing), chemical (desiccants and defoliants), or natural controls (such as reliance on unaided decomposition or decay processes). Another treatment strategy involves the use of fuelbreaks, or strategically located wide blocks or strips on which a cover of dense, heavy, or flammable vegetation has been permanently changed to one of lower fuel volume and reduced flammability (Green 1977; Agee et al. 2000). For example, the Civilian Conservation Corps constructed a massive network of fuelbreaks (100 to 300 m wide) in southern California to facilitate wildfire control during the 1930s.

More recently emphasis has broadened to encompass fuel manipulations, again primarily reduction, rearrangement, or removal, required to facilitate ecological restoration of fire to ecosystems from which fire was excluded during the twentieth century. Implicit in this latter aim is the desire to restore the biological diversity of forests to conditions that predate European settlement—that is, forests that are largely sustainable in the presence of periodic low-intensity fires.

It is important to distinguish fuel treatments for hazard reduction from ecological restoration, since fires have not been excluded from all ecosystems and because some forests possess a fuel structure and fire environment that supports high-severity fires. Although some Southwestern ponderosa pine areas have missed one to several cycles of low-intensity fires because of fire suppression policies, other fire regimes may not have been altered during the same period. For example, fuels have not accumulated in some shrublands, such as southern California chaparral, Great Basin sagebrush, or southwestern pinon-juniper forests. Furthermore, some forests (such as those at higher elevations in the Sierra Nevada or Rocky Mountains) may have been subjected to fire exclusionary practices, but not for a long enough time period to result in marked changes in anticipated fire severity. After all, these forests may have much longer fire return intervals than the period of fire exclusion to date, so fuel hazards may not be abnormally high.

Generally, the approaches taken for managing fuels depend on many factors, including the historical fire regime, the predominant vegetation, and land management objectives. For example, in low-severity fire regimes, such as Southwestern ponderosa pine, managers may aim to reduce the threat of crown fires by reducing fireline intensities while concurrently improving the ability of stands to withstand recurrent fires. Here the goal might be to create or restore fire-safe forests (Agee and Skinner 2003) by focusing on reductions to surface fuels, ladder fuels, and crown fuels.

Surface fuels (such as coarse woody debris, fuel jackpots, and understory vegetation) are removed to reduce the heat sources and flame lengths that carry a fire into tree crowns. Ladder fuels include low-hanging branches, small trees, shrubs, and lichens suspended from branches that bridge the surface and crown fuel strata. Removal of ladder fuels by pruning or low-intensity prescribed fires reduces the likelihood that flames from an eventual wildfire will reach into tree crowns or torch individual trees. Susceptibility to crown fire depends on foliar moisture content, surface fire intensity, and the height to the live crown base (Van Wagner 1977). So removal of fuel ladders effectively creates a vertical buffer to lessen the likelihood of a surface fire's climb into the forest canopy. Large trees should be retained to provide important diversity in stand structure, necessary for restoring forest stands to mimic pre-European settlement conditions. Species such as ponderosa pine generally have taller crown base heights and thicker bark, facilitating survival from low-intensity surface fires. Needle drop from large ponderosa pine trees also provides a porous mat of litter fuels that can carry periodic prescribed surface fires that maintain low fuel loads. Reducing tree crown density leaves a more open overstory, as overlapping tree crowns that touch one another are removed; these might support active or independent crown fires.

Most thinning operations remove small-diameter trees from the understory, otherwise known as "low thinning" or "thinning from below." Sometimes thinning removes overstory trees (that is, "canopy thinning"), although this may have less of an impact on reducing crown fire potential, especially if the heights to live crown base on residual trees are not lessened and if surface fuels are not reduced.

After thinning the untreated stand—for example, with machinery or chainsaws—the thinning slash and surface fuels are usually removed by piling and burning, since the thinned materials left on the site can constitute a dangerous fire hazard. Or the unwanted fuels can be removed physically, though often at great expense. If markets exist for the small-diameter trees that were cut as part of the thinning operation, those can be hauled away and processed to defray costs. Otherwise the thinned stand may be susceptible to a wildfire that sweeps through the area, wiping out not only the forest but also the investment in the thinning operation. The fire danger is highest when the needles dry out and remain attached to the fallen branches and limbs of the trees that were cut as part of the thinning operation. This so-called red slash fuelbed usually persists for a year or two following the thinning operation. During that time the stand's flammability can exceed that of the untreated stand because of higher surface fuel loads and greater fuel availability. If an ignition should occur, the fire's severity will exceed that of the untreated stand. Once slash disposal is performed, the area will require periodic maintenance, perhaps by low-intensity prescribed fires, to keep fuel levels at acceptable levels as the residual trees continue to grow.

Prescribed fire has long been touted as a useful tool for achieving a variety of resource management objectives. In fact, few fuel treatment alternatives seem so versatile as prescribed fire, especially considering the wide variety of possible uses, including (Fischer 2003):

1. disposal of logging residues and other woody debris;
2. removal of physical barriers to tree planting and travel by big game animals;
3. reduction of fire hazard by removal of or reducing fuel buildups;
4. wildfire control by burning out and backfiring to expand and strengthen firelines;
5. seedbed preparation for natural or artificial regeneration (exposing bare mineral soil);
6. competition reduction for tree seedlings by eliminating shrubs;
7. nutrient recycling in order to maintain site productivity;

8. site sanitation against insects and disease;
9. elimination of less desirable plant species;
10. maintenance of seral species on a site that would otherwise be taken by late successional tolerant species;
11. mimicry of a natural fire regime;
12. killing of invading trees and shrubs in order to maintain grasslands;
13. thinning of dense stands so that moisture and nutrients may become available to residual stems;
14. increase in distribution and amount of palatable forages for domestic and wild animals;
15. rejuvenation of decadent shrubs by top-killing, in order to induce sprouting and thus improve deer and elk forage;
16. maintenance or restoration of "natural" conditions, such as in a national park or wilderness area.

Prescribed fire can be used for such a large variety of objectives in part because effects can be varied depending on treatment objectives and selection of the appropriate fire prescription window, the conditions under which the fire is ignited. Broadcast burning is applied when the intent is to consume fuels with a relatively intense fire applied over a large area. Understory burning uses a low-severity fire applied with the intent of minimizing damage to standing trees. Stand replacement burning may be used to intentionally kill the overstory trees. Prescribed fire effects can be varied depending on type of fuel, moisture, wind speed, and firing method (pattern and ignition device). For example, a handheld drip torch or fusee will produce a lower severity fire over a more restricted area than is produced by mass ignition by helicopter.

A variety of firing techniques have been developed for implementing prescribed fire objectives. Head fires produce rapid, high-intensity burning pushed by the wind or slope, desirable for treating large areas, brush fields, or clear-cut areas (Martin 1976). Initially backed against the wind or slope to create a safe buffer strip, a head fire is then ignited and allowed to burn with the wind until the buffer is encountered and the perimeter secured.

A backfire, by contrast, produces a slow, low-intensity burn, most appropriate for burning under a tree canopy, in heavy fuels near a fire control line, or downhill (ibid., p. 149). Ignition takes place downwind, and the fire backs slowly into the wind. A strip head fire may be appropriate as an alternative to the head fire. After the safe buffer is created, successive strips are ignited with fire intensity adjusted by the strip width (ibid., p. 149). If access within the burn unit is a problem, a helicopter may be used to treat the area with spot head fires, similar in effect to the strip head fire.

Alternatively, flank firing or center (or ring) firing may be used, depending on the objectives and personnel. Multiple flank fires are ignited against the wind and are allowed to spread laterally into each other, with fire intensity determined by the speed of ignition and amount of fire laid down by igniters. After securing the safe buffer, the several burners progress into the wind, each adjusting ignition speed in order to produce desired flaming. This technique might be appropriate for light fuels under a canopy. With center firing, the unit center is ignited first, followed by successive rings that are drawn into the draft created by the center fire. This is a rapid technique with good air dispersal, but dangerous convection columns may develop and require a large crew (ibid., p. 149).

The biggest shortcomings of prescribed fire are the risks of escape and smoke. Even so, when judiciously applied, prescribed fire provides a low-cost method for treating fuels and meeting other resource objectives. Ultimately, forests that are safe for prescribed fire should be able to withstand wildfires under most circumstances.

Fuel Treatment Effectiveness

Treatments such as prescribed fire and mechanical thinning reduce, remove, or rearrange flammable biomass but still leave behind materials that can burn—especially during a drought, when fuels are generally more available. Prescribed fire techniques and objectives were addressed in the previous section. Thinning is a silvicultural technique for manipulating the density and composition of a forest stand to achieve management objectives. Thinning with slash removal reduces flammable biomass that could be available to fuel wildfire rate of spread and intensity.

Conventional wisdom might suggest that fuel treatments would be less effective under drought conditions, since the residual fuels might become more flammable. Although fuel modification will not lessen wildfire spread and severity in all circumstances, there are also many instances in which practitioners have observed a raging crown fire dropping to the surface when it encounters an area where fuels have been treated (that is, at the edge of a fuelbreak). To date, much of the evidence for fuel treatment effectiveness has been isolated and anecdotal—and often controversial.

The progression of the 2002 Hayman fire, the largest fire in Colorado history, is illustrative of some of the controversy. Occurring in the midst of an unprecedented drought, high southwest winds pushed the fire into a huge, two-pronged fire run on June 9, racing through several areas in which fuels had previously been modified by timber harvest. In fact, the June 9 run covered an area (20,000 ha) almost half the total size of the Hayman fire (55,500 ha)—all in just one afternoon of burning. Recent wildfire and prescribed burn areas effectively pinched off one of the fire prongs; other fuel modifications encountered after winds shifted seemed to marginally affect burn patterns, until the time of eventual fire containment on June 18 (Graham 2003).

So the Hayman fire illustrated that some fuel modifications may not stop or retard a fire's spread, especially when inspired by drought and high winds, although the fire perimeter had been confined by fuel modification areas encountered later in the fire's life (that is, June 18–19). Also, the previous wildfire and prescribed burn appear to have dampened fire spread. The timber harvest areas overrun during the fire run on June 8 were not carried out with the intent of reducing fuel hazards, which may partially explain their minimal impact on the overall severity and size of the major fire run—along with the drought and winds.

Generally fuel treatments are not designed to stop a fast-moving fire in its tracks, especially if firefighters are unable to get into the treated area and make a stand against the fire—say, with the assistance of air tanker retardant drops. With the Hayman fire, the explosive fire run occurred so quickly and conditions were so hazardous that no firefighters were placed in harm's way during the major run on June 8. In fact, fires can be expected to burn through areas where fuels have been treated, albeit with lower spread rates and intensity. Furthermore, although treated areas may not stop a fire, a well-designed system of fuel treatments will allow safe access and egress for firefighters, as well as provide an opportunity to assess whether other strategies, such as lighting a backfire, might be warranted if justified by safety and environmental considerations.

Fuel treatments in the United States have been studied since the 1950s, although quantitative evidence for the performance of fuel manipulated areas during wildfires is limited to relatively few case studies (Omi and Martinson 2002, 2004). Even so, the 2000, 2002, and 2003 fire seasons have spurred interest in a national expansion of fuel treatments. In particular, the 2003 Healthy Forests Restoration Act of the G. W. Bush administration provides for accelerated rates of commercial logging, mechanical thinning, and prescribed fire in Western forests (Chapter 4). That expansion is extremely controversial, not only because of the increased cutting in national forests but also because treatments may be carried out in ecosystems that are not in need of restoration. Furthermore, road-building in remote areas also will be required and permitted. Environmentalists have traditionally been extremely suspicious of both logging and road-building, especially in remote or pristine areas. In addition, the scale of the proposed expansions in fuel treatments is unprecedented, even though most experts acknowledge that reductions in the frequency and severity of wildfires may be possible only through landscape-scale treatments. Unfortunately, few studies have been published on the effectiveness of fuel treatments carried out at landscape scales, so little information is available for assessing the likelihood of success for the scale of fuel modifications proposed under the mandates of the 2003 Healthy Forests Restoration Act. Adding to the controversy has been an apparent withholding of funding to carry out restoration projects.

To date, most fuel treatment projects have focused on relatively isolated stands, mostly scattered with little connectivity across a landscape—with the possible exception of fuelbreak and firebreak systems such as those installed by Civilian Conservation Corps crews in southern California during the 1930s. Thus to date most of the evidence for and against fuel treatment effectiveness comes from relatively few studies that compare wildfire severity in adjacent treated versus untreated stands. For example, Omi and Martinson (2002) and Pollet and Omi (2002) have studied differences in three indicators of wildfire severity (crown scorch, stand damage, and ground char) in untreated versus treated stands in eight wildfires occurring between 1994 and 2000. Their results confirmed observed reductions in wildfire severity from treatment in all burns studied (Omi and Martinson 2003).

While not providing conclusive proof on the efficacy of fuel treatments, these studies suggest that wildfire severity can be reduced in treated areas, especially under the severe burning conditions that characterized the fires studied. Furthermore, the effectiveness of treatments will depend on many other factors, such as treatment type, time elapsed since treatment, and the standards to which treatments are implemented. Nonetheless, such wildfire studies represent an ongoing effort to fill the void of information on the effectiveness of fuel treatments.

Managers for years have attempted to reduce wildfire hazards by treatments that produce structural changes in the forest, such as reduction in tree density and surface fuels using silvicultural treatments such as thinning and prescribed fire. As an alternative to studying the effects of various fuel treatments on actual wildfires, an experimental approach known as the "fire and fire surrogate study," has established replicated treatment sites across the United States, all in low-severity fire regimes. Investigators are interested not only in possible effects on wildfire severity but also in the possible effects on such ecological attributes as nutrient cycling, seed scarification, plant diversity, disease and insect abundance, and wildlife habitat (http://www.fs.fed.us/ffs/). Although still in progress, this study will add greatly to our knowledge about the effects of treatments that attempt to mimic the ecological functions of low-intensity fires by thinning the understory vegetation and fuels within a forest.

Judicious fuels management can mitigate wildfire severity in a variety of ecosystems, but even treated areas may burn with high severity under extreme environmental conditions. To be most effective, treated areas need to be placed in strategic locations across a landscape, in areas where firefighters can take a stand (Finney 2001). Furthermore, the treatment must fit the ecological conditions in the area. Omi and Martinson (2004) note that fuel treatments may be less effective in ecosystems characterized historically by infrequent, high-severity fires—because of the extreme environmental conditions necessary to permit fire spread in the first place. Thus high elevation–high latitude forests may have experienced little change in fuel profiles as a result of twentieth-century fire exclusion, so fire severity was and remains high to extreme during fire outbreaks. For example, Alexander et al. (2001) suggest that fire intensity may increase in boreal forests of the Canadian Northwest Territories after fuel treatment. In such areas, openings in the forest canopy also may expose surface fuels to increased solar radiation, thus lowering fuel moisture and possibly promoting production of fine herbaceous materials. Forest openings thus created also may be exposed to intensified wind vectors, accelerating both fuel desiccation and heat transfer.

Additional information is needed on the ecological differences between thinning versus prescribed fire, including the duration of each treatment's effectiveness. In particular, the case studies noted above hint at the possibility of qualitative differences between the two treatments, suggesting that areas that have been intentionally burned may possibly be safer than areas that have only been thinned. Although thinning and prescribed fire both remove small-diameter trees, prescribed fire may offer possible advantages in terms of the removal of fine fuels, albeit patchily throughout a stand, both horizontally and vertically. It is also possible that prescribed fire may "harden" a stand against future wildfires in ways that thinning alone will not. This possibility deserves attention in future research.

Fire specialists have long recognized that fuels are the only part of the fire environment that can be managed effectively. Managers treat fuels to alter the arrangement, size, and flammability of fuel profiles, usually to reduce the area burned and the severity of subsequent wildfires, to enhance firefighter safety, or to reduce fire control costs. Other objectives may be achieved, including wildlife habitat enhancement, removal of exotic plants, or the achievement of a desired composition of species in an area. So it will be extremely interesting to see if initiatives (such as the National Fire Plan and Healthy Forests Restoration Act) are successful in reducing wildfire costs and losses over the long term. Only time will tell, especially since so much area is in need of treatment in the western United States. In the short term, large areas may continue to burn until existing backlogs of areas requiring treatment are brought under management. However, based on the limited but growing body of knowledge, it appears that increased emphasis on fuel modification should eventually provide a good alternative to the costly, all-out suppression policies for managing fire on the landscape.

Wildfires in the twentieth and early twenty-first centuries have demonstrated clearly the futility of attempts to eliminate fire from natural landscapes. Fuel treatments will possibly ease the tensions created between our society and forest fires, but only if people learn to live with the notion that our wild areas will burn. In some respects, fuel treatments provide us with a choice between coexisting with out-of-control conflagrations and with manageable ignitions, although the choice is not always clearly defined.

Commercial Logging and Wildfires

Increases in commercial timber harvest have been proposed as another means for managing fuels and controlling wildfire outbreaks, although no strong consensus exists on that possibility (Omi and Martinson 2004). For example, Weatherspoon and Skinner (1995) found fire severity to be greater in regenerated clearcuts of northern California fir forests than in unharvested stands. By contrast, Omi and Kalabokidis (1991) observed higher fire severity in unmanaged lodgepole pine forests than in intensively managed Targhee National Forest lands within the greater Yellowstone area.

Commercial timber harvesting can reduce heavy fuels and fuel ladders in an area but may also increase fire threats unless the slash (residual tree tops, limbs, branches, and other coarse woody debris) is properly removed or reduced. Also, ambiguities about the relationship between timber harvest and wildfire are understandable, especially since historical records on harvest levels and wildfire burned area show little relationship (Gorte 2000). Actually, the data in this study may be too coarse for the desired comparison, since they do not represent the actual wildfire area burned in areas that were harvested for timber during the period 1986–1999. Also, the data aren't available to show which harvests, if any, were carried out for the purpose of mitigating the size of wildfire area burned. Even so, Gorte (ibid.) illustrates the dramatic drop-off in timber harvests since the 1980s, as well as showing how burned area may fluctuate dramatically regardless of harvest volumes, responding instead to fuel, weather, and topographic variations wherever the fires occurred.

In a subsequent report, Boxall (2002) notes that since 1945, timber harvest levels had steadily increased approximately to the 1988 levels through the early to mid-1970s before dropping to the 1982 levels shown in Gorte (2000). During roughly the same time period (that is, 1945–1980), national forest annual burned areas stayed relatively low. One inference from this extended analysis is that the severe fire seasons since 1988 may, in fact, be a legacy of earlier accelerating timber harvests from 1945 to 1975, and not the drop-off in logging during the 1990s.

In reality, the relationship between logging and wildfire is not well defined, and may depend as well on harvest practices employed, droughts and climate change, expansions in the urban interface, and fuel accumulations from twentieth-century fire exclusionary practices. What seems clear is that, like thinning, fire hazards can be increased by harvest practices that increase surface fuels, ladder fuels, and crown densities while removing all large trees. In such areas, the risks and hazards caused by wildfire may increase after logging. On the other hand, logging practices that decrease surface fuels, ladder fuels, and crown densities while leaving larger trees should result in forests that are more resilient to fire.

Relationships between timber harvest and wildfire area burned will become increasingly important as the 2003 Healthy Forests Restoration Act is implemented. It will be especially instructive to monitor those burned areas where harvests have been conducted, as well as the severity with which fires spread through the harvested (and unharvested) areas. Equally important will be records on the stand structure in treated versus untreated areas, as well as the extent to which fuel hazards were mitigated by the harvest and subsequent treatments. In the meantime, the relationships between commercial logging and fire will remain ambiguous and controversial.

Wildland Fire Use

Although large fires receive a lot of attention because of their economic consequences (costs and losses), we should not lose sight of the fact that extensive burns also produce significant ecological consequences that may drive important ecosystem processes, including many of the benefits attributable to prescribed fires. In fact, large wildfires may reduce or remove fuel hazards that may be important in the next fire outbreak.

Wildland fire use is the current lexicon for natural ignitions that are allowed to burn, with management intervention if needed, in order to achieve resource benefits. The goal of the wildland fire use program is to allow lightning ignitions to resume a more natural role in ecosystems in which fire has been important. Earlier vestiges of this program included the so-called let burn, natural fire management, or prescribed natural fire policies of federal agencies.

The wildland fire use program and its earlier precursors have survived several significant disastrous escapes that threatened to end the program. Most notably, the 1988 Yellowstone fires prompted a thorough policy review but mostly re-enforced the importance of programs that permit lightning ignitions to assume a more natural role in ecosystem management and restoration. Several improvements were made to implementation procedures, mostly aimed at reducing the uncertainty of eventual fire size and effects, mitigating risks of fire escape, and strengthening accountability.

Wildland fires managed for resource benefit are lightning-caused fires that are allowed to burn, usually within an identified, undeveloped management area that has undergone extensive planning. These fires are monitored and evaluated according to a predetermined schedule that is revalidated daily. Suppression actions are taken if these fires demonstrate behavior that is inconsistent with resource management objectives or if the fire approaches its predetermined boundary. Not all fires grow large—some may be extinguished by rain or snow or otherwise confined to small areas. However, the potential always exists that a fire may grow to large size, especially if the ignition occurs early in the fire season. Once the flaming area becomes large, fire managers may have few alternatives if weather conditions shift toward the promotion of extreme severity burning conditions. The passage of a series of dry, cold-front systems in August during the 1988 Yellowstone fires (Fuller 1991) provide an illustrative example of this type of weather shift and the subsequent futility of fire control efforts.

USDA Forest Service fire planners assess risk and predicted fire behavior and growth, plan for contingencies, and determine the maximum limits of the fire area (also called the maximum manageable area, or MMA). The RERAP (Rare Event Risk Assessment Process, see Chapter 7) provides a tool that is used to decide on the wisdom of allowing a fire to burn, based on the likelihood of season-ending precipitation before valued resources (such as human communities) are threatened. Planners must also define the thresholds or trigger points that may signal the need to mitigate threats, including fireline construction, helicopter bucket or air tanker retardant drops, or firing operations.

The National Park Service has been a leader in natural fire management policies since 1968, with the most aggressive programs in Yosemite and Sequoia–Kings Canyon national parks. More than 80 to 85 percent of the total area in Yosemite and Sequoia–Kings Canyon has been designated as Resource Benefit zones, where natural fires will be permitted to burn. Yosemite has been experimenting with prescribed natural fire in the Illilouette

Creek basin since 1974. During that time, fire scientists have noted a pattern emerging on the landscape whereby the size and intensity of new fires is limited by recent, nearby burns. New fires run out of fuel and go out, or at least burn with reduced severity (Van de Water n.d.).

The historical role of fire must be established and considered before wildland fire use zones are established. Fuel loads and predicted fire behavior must be within prescribed ranges, and plans must address risk of escape, safety, tactical suppression needs, and smoke management concerns.

Van de Water (ibid.) shows that 523 fires in Yosemite burned almost 25,000 ha (more than 60,000 acres) under wildland fire use guidelines between 1972 and 1999. In some years few fires or no fires were allowed under these guidelines, while the greatest number (forty-seven) occurred in 1987. In 1988, slightly fewer fires (forty-four) burned the largest annual area (5,109 ha). The overall burned area in any given year may be small, but numerous constraints restrict the size of areas treated, not the least of which are manager reluctance or fear of job loss (van Wagtendonk 1995).

More important, park scientists monitor and evaluate the effects from each fire and keep track of the area burned compared with historical fire rotations in different vegetation types. These measures provide insight into the biological effects of wildland fires and also indicate the progress of fire restoration programs. Overall, the annual area burned under wildland fire use programs tends to lag behind the total area required to ensure timely restoration of fire-adapted ecosystems. However, the wildland fire use program will continue to be an essential tool for managing fires, especially if complemented with strategic fuel modification efforts in key locations and boundary zones.

Fire Suppression

Fire suppression activities draw the most attention and might be the best-understood part of systematic fire management, at least by the general public. News media coverage has made air tanker retardant drops and firefighters adorned in yellow Nomex attire a familiar staple of every fire season. Most people understand the mechanics of dousing a flame, extinguishing embers in a campfire, or blowing out a candle. The same general principles apply to firefighting. Firefighting can be tedious, hard work, yet dangerous. But the feelings of accomplishment and camaraderie appeal to many, including crewmembers and overhead teams that deal with large fires.

A fire that is discovered or reported to a local jurisdiction will set in motion a series of planned and organized responses. Typically, a fire dispatcher familiar with the local terrain and values-at-risk will call upon appropriate suppression resources to respond to the incident. If additional information is needed, a local official may drive to or fly over the fire to size up the local vegetation, burning conditions, and potential problem areas. Depending on the present and projected levels of fire danger, dispatched resources might include one or more hand-crews, fire engines, or perhaps an air tanker on standby at a regional airfield. Particularly challenging fires may dictate the need for skilled and experienced hot-shot crews. In remote, mountainous terrain, smokejumpers may parachute down to a fire, or a helicopter crew (also known as a helitack crew) may be dispatched and rappel to the fire. As a fire grows in size, or if other ignitions occur, additional regional or national resources may be mobilized, including an administrative or supervisory team known as fire overhead. If the fire exceeds the capabilities of the local agency with jurisdiction over the fire, that agency can request additional resources through the regional Geographical Area Coordination Center (GACC) or from the National Incident Coordination Center (NICC) in Boise, Idaho. As a fire grows in size and complexity, increasingly sophisticated fire organizations will be required. Thus a Type II overhead team may be dispatched initially to assist local jurisdictions in managing an incident. Eventually a Type I team may be dispatched if fire complexity exceeds the capabilities and resources of the Type II team.

Not all suppressed fires are subjected to aggressive initial attack. Fire management plans are written to distinguish areas in which fires will be fought aggressively from sensitive zones (such as wilderness areas) where fires will be managed under minimal impact suppression guidelines—that is, nonmechanized equipment and less reliance on tree cuttings or other activities that will disturb the postfire landscape. Also, firefighter safety considerations may sometimes dictate the need for less aggressive suppression strategies. For example, instead of confronting a raging inferno in a deeply dissected canyon, firefighters will back off and take a stand on a safer ridge line.

Initial Attack

All fires to which firefighting forces respond undergo an initial attack, or the first suppression activity on an incident. The first forces on a fire (typically a fire engine or helitack crew) will perform the initial attack under supervision of the incident commander (IC). The IC, usually the most experienced firefighter on the scene with leadership capabilities, will size up the fire before deciding upon strategy and tactics to be employed in managing the fire safely. Strategies may include direct attack on the fire's edge, beginning on the flanks in order to minimize the fire perimeter, or indirect or parallel attack at a distance from the fire's edge. Tactics might include construction of a fireline along the flanks of the fire, incorporating natural barriers as encountered.

The number and type of resources involved in initial attack will depend on the fire danger, fuel types, values at risk, and land management objectives, among other considerations. These same considerations, along with the fire's behavior, will dictate the width of fireline constructed or type of hose lays used on the fire. Generally, the resources involved will be small in number (as the fire size is relatively small) and may vary from one to several single resources (for example, an engine or several crews). However, the IC may request additional support from the dispatcher, such as crews equipped with hand tools or air tankers with retardant. Usually, the fire will be contained within a single work period (less than twenty-four hours), although mop-up may extend over several days.

The mechanics of firefighting are fairly straightforward, involving activities that essentially break a leg on the fire triangle, as noted previously (see Forest Combustion, pp. 114–119). The most common strategy for controlling a fire involves construction of a fireline, cleared to bare mineral soil to remove fuel in the fire's path. The intention is to completely surround the fire and any isolated spot fires apart from the main fire body. Once constructed, the line must be held by burning out unburned fuel islands between the line and the fire's edge, or by cooling off and mopping up hotspots.

Close to 95 percent or more of all fires are successfully controlled by initial attack crews, usually by construction of a fireline around the fire perimeter followed by patrolling and mopping up of hotspots. Some lightning ignitions may go undetected and burn out on their own if the thunderstorm is accompanied with sufficient moisture or if the ignition has occurred in a remote area with low values-at-risk. Firefighters enjoy greatest success when fires are attacked during the initial phases, before the fire reaches flashy fuels or encounters high winds that can produce high spread rates and breach control lines. Early attack also may preclude the formation of an organizing convection column that itself will spur fire growth through the creation of updrafts and downdrafts that can propel fire spread beyond control lines. Even so, constant vigilance is required even after a fire has been completely encircled with constructed firelines or isolated by natural barriers, to ensure that an errant wind gust does not propel glowing embers across the line. Holding a fireline is greatly facilitated by the availability of water to douse flames and hotspots.

Extended Attack, Project Fires, and Megafires

Fires that are not contained or controlled by initial attack require an extended attack, involving additional firefighting resources ordered and supervised by the IC in conjunction with the dispatch office. A fire that is designated as an extended attack incident has essentially exceeded the control capabilities of the initial attack forces, an admission that the fire could transition to a much larger and more costly incident. The IC might need to shift thinking from keeping the fire as small as possible to a realization that the fire behavior is simply too extreme to risk putting firefighters in harm's way. If the incident continues to grow in size and complexity and still seems uncontrollable, the fire may be called a project fire, or campaign fire, requiring perhaps a more complex organization (Type II or Type I overhead team); a fire camp may be necessary for feeding, lodging, and equipping firefighters from a variety of locations for a time lasting anywhere from one to several weeks. Megafires require even more people, more equipment, and greater commitment of financial resources.

Williams (2003) suggests that the extended attack fire and the megafire are the two most important and challenging fires to suppress—one because of safety and the other in terms of exorbitant cost. Seventy percent of fatalities occur on extended attack fires, usually as the fire transitions from an attackable fire to one that should not be confronted. Extended attack fires typically occur during periods of high fire danger, when suppression crews with slim supervisory oversight are fatigued and overmatched by the expanding fire. Although typically representing only 1 percent of all ignitions, large project and megafires may account for as much as 80 percent of suppression costs and 90 percent of annual burned area. The best response to megafires may be to wait for favorable weather, rather than attempting an aggressive attack.

For the larger fires, firefighters may come from federal agencies, native tribes, state forestry agencies, or low-security detention facilities; other labor sources include local volunteers, private fire departments, or contractors. Seasonal employees with nonfire job responsibilities may be assembled and organized into Type II crews. The military (Army or National Guard) may be mobilized on the largest or most complex incidents. Although the strategies for fighting a fire may be similar as the incident grows in size and complexity, the tactics may differ considerably, relying on bulldozers and interregional hot-shot crews for fireline construction on "hot" sectors of large fires.

Large fire management requires a paramilitary organization. The growing size and complexity of the fire will dictate that the IC delegates some of the responsibilities attendant with managing the fire to other support staff. The Incident Command System was developed in the 1970s to provide a standardized framework (including terminology) that could be applied across multiple agencies for managing any emergency of any size. Today, the ICS has become a system that is admired and imitated by emergency response organizations globally. In fact, the system used by federal agencies with fire management responsibilities has become a model for responding to incidents as diverse as the *Exxon Valdeez* oil spill, the *Columbia* shuttle recovery, and the 911 terrorist bombings. In fact, the USFS sent two Type I teams to assist with coordination, rescue, and salvage operations for the World Trade Center cleanup.

The ICS organization develops around five major functions that are required for managing any incident: Command, Operations, Plans, Logistics, and Finance/Administration. On most incidents the command function is handled by a single IC, who determines incident objectives and strategy, sets immediate priorities, and establishes an appropriate organization. Tactical operations are managed by the operations section chief. The planning section chief supervises the collection, evaluation, processing, and dissemination of incident action plans. The logistics section chief supervises all incident support needs. The finance/administration section chief supervises all financial aspects of an incident.

On a small incident, the IC may be able to perform all the major functions—that is, overseeing operations, plans, logistics, and finance. However, the ICS was designed to handle any type of emergency incident, regardless of size or complexity, by adding support positions to each major functional activity. Thus each major function identified above may have a support staff that expands with the complexity of the incident. For example, Figure 5–16 shows a more complex organizational structure that might be appropriate for a project fire with significant air and ground resources serviced by a fire camp and off-site aerial bases. The IC might use additional staff, such as an information officer, liaison officer, and safety officer. The information officer would be responsible to the IC for communicating with the news media, to local publics, to incident personnel, and to other agencies or organizations. For multijurisdictional fires that involve several agencies, a liaison officer might be required for keeping involved agencies apprised of incident status, or possibly, for communicating with crews consisting of prison inmates or non-English speakers. The safety officer would be responsible for recommending to the IC measures involving personnel safety, for identifying potentially dangerous situations, and for investigating accidents that might occur.

Likewise, each staff position (operations, plans, logistics, and finance/administration) may be supported by additional branches and units necessary for managing the incident (Figure 5–16). This type of organizational structure also lends itself to the many activities that go on "behind the scenes" in a project fire camp. For example, the daily incident action plan detailing the operations to be carried out on the day and night shifts may be written as a collaborative effort between the operations and planning section chiefs. Also, the entire team may provide input to the Wildland Fire Situation Analysis (WFSA), describing the complexity of an incident and documenting the processes used to arrive at suppression decisions. Alternatively, on some complex incidents, the WFSA may be prepared by a team brought in from the outside precisely for that purpose.

Firefighting Tools

Regardless of the organizational complexity of the incident, firefighters on the ground are required to carry out and complete suppression of any threatening fire. Even as technologies for

Figure 5-16
Fire Organization for a Large Incident
A complex incident will require support position for each major function, including the Incident Commander.

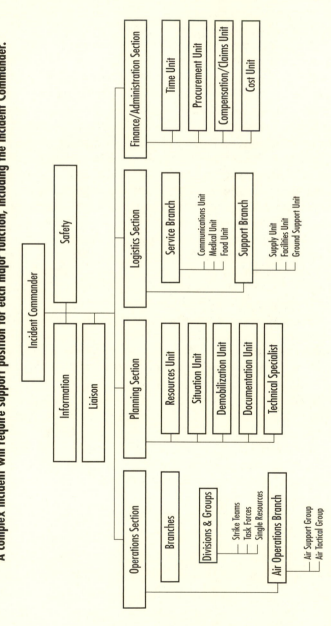

Source: www.nifc.gov/fireinfo/ics_disc.html.

detecting, mapping, and monitoring fires become increasingly sophisticated—for example, involving satellite or remote links to
computers in fire camp—the brunt of the firefighting effort still
relies on people on the ground chasing smokes and making sure
that the fire is out.

Early in the twentieth century, the primary tools of firefighting included shovels, axes, and handsaws. By the mid-twentieth
century, chainsaws, motorized vehicles (including tractors and
bulldozers), and compressed-air water pumps had improved firefighter capabilities. And today the efforts on the ground are augmented by air tankers, helicopters, and sophisticated fire engines—each taking advantage of increasingly advanced
technologies. Although the arsenal has expanded considerably
since the early days, the basic tools of the trade haven't changed
much; firefighters still rely on shovels, axes, and saws to build
line and put the fire out. Basic tools employed on the fireline include the following:

> Shovel: A combination tool (modified from the familiar
> garden variety) with a cutting edge (for chopping down
> small trees, and cutting limbs and roots) and angled for
> scraping/digging. Useful for scraping away needles, duff,
> and litter while constructing and clearing a fireline to bare
> mineral soil. Shovels also are useful for throwing dirt at
> hotspots.

> Pulaski: A combination tool (ax and mattock, or grubbing
> head). Used to cut trees and lop limbs (ax side) and to
> dig/scrape (mattock side).

> McLeod: A combination tool with heavy-duty rake and
> sharpened hoe. Used to cut deep, matted litter (hoe) and
> clear loose surface debris (rake).

> Chainsaw: Straight or bow-bar for cutting trees and limbs.
> Firefighters use smaller chainsaws for cutting trees, shrubs,
> and downed logs to clear a fireline. Fellers use larger
> chainsaws for cutting down large-diameter burning trees
> and snags.

> Backpack pump: Collapsible neoprene backpacks filled
> with water (5 gallon or 18.9 liter capacity), connected to
> small hose and nozzle. Useful for spraying water during
> mop-up operations.

Drip torch: Canister with piloted ignition filled with gas-diesel mixture. Used with fusees or other ignition devices to burn out or backfire fuels between established control lines and the main fire.

Brush hook: A tool for cutting limbs and branches, primarily in shrub fuelbeds. Although it is primarily a cutting tool, the back of the hook can be used for scraping soils; used primarily in southern California.

Primacord: Blasting cord used for clearing fuels to mineral soil for constructing fireline. Used in remote areas; requires personnel qualified to work with explosives.

Fire engines and heavy equipment, such as bulldozers, augment crew line-building activities. Engines carry a supply of water, tools, hoses, and pumping equipment for delivering water to a fire. They also can draft water from nearby rivers or lakes for replenishing water supply. Bulldozers push, scrape, and pile debris as part of fireline construction and also create access routes through heavy fuels for engines and crews working the fire. In the South, tractor plows are used to furrow a line around a fire.

Numerous aircraft support fire operations on the ground, although most fires do not require aerial assistance. Every year, thousands of initial attack fires are suppressed without the need of aerial support. Others may be suppressed without aircraft because of flight distance or unavailability. Even so, an airtanker dropping retardant may provide the most familiar image of firefighting to most people. The aircraft used in fire operations include the following types.

Helicopters: Provide support for firefighters working a fire perimeter cooling hotspots; can slow a fire spread by dropping water, foam, or retardant on burning trees, shrubs, and structures. Large helicopters carry up to 2,000 gallons (7,571 liters), though medium and light helicopters have correspondingly lower capacities, down to 300 gallons (1,136 liters) or less. Helitack crews can rappel down to fires in remote areas. Also indispensable for conducting large-scale firing operations on wild and prescribed fires.

Air tankers: Large planes equipped with tanks for transporting and dropping fire retardant (for example, diammonium phosphate) or water in order to slow a fire down in support of fireline construction operations on the ground; primarily for use in initial attack and structure protection support. Typically, a red dye is added to chemical retardants to provide pilots with visual markers of drop accuracy. Capacities range from 2,000 to 3,000 gallons (approximately 7,571 to 11,356 liters). Single-engine air tankers have less capacity but greater mobility.

Lead planes: Smaller, mobile airplanes used to provide reconnaissance of burning areas from above and guide air tankers to drop locations.

Infrared reconnaissance planes: Small airplanes equipped with specialized infrared mapping systems for detecting hotspots inside and outside a fire perimeter. Sensitive scanners are able to detect heat with a high degree of accuracy (that is, a 6-inch [15-cm] spot from 8,000 ft [2,438 m] elevation). Geographic referencing of hotspots via global positioning systems provides useful reconnaissance information.

In the United States, many aircraft are either surplus from the military or date from World War II, including helicopters, air tankers, modular airborne fire fighting systems (MAFFS), lead planes, and infrared aircraft. Most airplanes are leased by governmental agencies from private contractors who convert former military aircraft to carry and deliver fire retardant. The aging U.S. air tanker fleet (forty-four ships in 2002) has caused great alarm, especially with two fatality accidents in 2002, involving planes that broke apart in the air while delivering retardant to fires. The planes involved in crashes that year included a World War II-era PB4Y-2 that broke apart near Estes Park, Colorado, in July 2002 and a C-130A air tanker whose wings separated from the fuselage in June 2002 near the town of Walker in eastern California. The planes were fifty-seven and forty-six years old, respectively. Fatigue cracks were found in the wings of both planes by investigators from the National Transportation Safety Board (NTSB). At one point, following the second crash, the entire USDA Forest Service fleet was grounded; later, the government permanently banned the use of the PB4Y-2 and C-130A aircraft for firefighting use.

In May 2004 the federal government decided to terminate the contract for the thirty-three fixed-wing large air tankers still operating within the firefighting fleet, after the NTSB released a report of findings following the fatality accidents. The report found no effective mechanism for ensuring the continuing airworthiness and safety of these older ships, concluding that the older aircraft posed an unacceptable risk to contract aviators, firefighters, and the general public. The termination of the contract essentially ended a colorful (and controversial) chapter in the history of firefighting in the United States. Earlier the U.S. Forest Service had discontinued use of eleven Beechcraft Baron aircraft owned by the agency and used as lead planes. The May 2004 contract cancellation did not apply to the eighty single-engine air tankers and eight military C-130 E and H model aircraft equipped with MAFFs, whose airworthiness is the responsibility of the military.

Air tankers seem to attract controversy because of their high visibility and cost. Cynics speak of "political retardant" in reference to planes that are called into action when they are not needed, appealing instead to influential politicians and a public that wants to see agencies taking action against a fire. On the other hand, some believe that the 2003 Cedar fire in southern California escaped control in early evening because of the unavailability (the result of concerns for pilot safety) of air tankers stationed in the vicinity. The truth is that on some fires, air tankers are extremely useful in helping firefighters on the ground to gain the upper hand on a wildfire, while on others their excessive use may be a waste of money.

In some locations (particularly near large bodies of water), U.S. agencies have increasingly relied on the Canadair CL 215 and the larger CL 415 aircraft. These planes offer advantages because of their ability to scoop water from a lake or sea for subsequent drops on land. Also, airworthiness is documented and supported by Canadair, a company based in Canada.

In 2004, tests were carried out on a Boeing 747 jumbo jet, retrofitted for firefighting by the Evergreen International Aviation Company. The jet will likely have limited usefulness on small fires and won't be able to navigate close to fires in deep canyons—but it can carry a 20,000-gallon (75,708-liter) payload, which may be useful on large fires. As of this writing its potential usefulness and cost are unknown, and it has yet to earn approval by the Federal Aviation Administration.

The retardant dropped by air tankers and helicopters (sometimes called slurry) consists of water (85 percent), diammonium phosphate or sulfates (10 percent), and minor ingredients (5 percent). The mixture weighs about 9 pounds per gallon, or approximately 1 kg per liter. Diammonium phosphate is the active ingredient that inhibits combustion and fertilizes the soil after the fire is out. Minor ingredients include iron oxide (a red dye, allowing the pilot to check the accuracy of the drop), and gums and clay thickeners.

Helicopters used on fires include the heavy-lift Erickson helitanker, military-style UH-1 Hueys, Kaman-built K-Maxes, Bell Rangers, several Sikorski models, and Boeing-built Chinooks. Payloads vary from 600- to 1000-gallon (2,271 to 3,785-liter) buckets filled from tanker trucks or water springs to the 2,600-gallon (9,842-liter) detachable tank on the helitanker that can be filled in 45 seconds from a lake or deep hole in a river (Enders 2001).

Fire engines on the ground make use of water and water enhancers such as fire retardants and foams. Water has the capacity to absorb and carry away heat, even as it is converted into steam. However, water's strong surface tension causes it to bead up and roll off most fuels too fast to be able to absorb its full heat capacity (Astaris 1998). Retardants and foams make water more efficient as a suppressant by spreading water out over a surface and penetrating more deeply into porous fuels, much as a detergent extends water's usefulness for cleaning purposes.

In the United States, much of the equipment and many of the services used on ongoing fire incidents are contracted by the federal government, including aircraft and aircraft support services, but also commissary, retardant, showers/portable toilets, and engines/crews, to name just a few. Other contracted services may include bulldozers, water tenders, flatbed trucks, and chainsaws/felling crews. In addition, the Federal Excess Personal Property (FEPP) program allows a federal agency to acquire title to government property (that is, purchased for use by a federal agency but no longer needed by the original purchaser) for loan to one of the fifty states for use in wildland fire protection. This can include trucks, aircraft, personal protective equipment, motor oil, nuts and bolts, and fire hose (but not real estate).

Repeated heavy fire seasons have spawned a huge market for support services, such as full-service kitchens, portable toilets, tools, commercial retardants and foams, and contract crews. The breadth of this market becomes evident especially on project and megafires, for which all the equipment and supplies are shipped to fire camp and employed to manage the incident. The service market for wildland fire use or prescribed fire programs is much less developed at this time.

Communications

Good communications are essential for carrying out fire management activities, whether in conducting fire suppression operations or in establishing two-way dialogues with the public. Firefighting is carried out with some of the trappings of a paramilitary campaign, with a unified command and control structure that requires a good communications infrastructure. Orders given by supervisors must be clearly understood and carried out—or else the system breaks down. Communication failures are often at the root of mistakes, and in the worst case, can lead to fatalities. Fortunately, the Incident Command System (ICS) adopted by the agencies has been developed to facilitate improved communications by adoption of standard terminology and chain of command.

Communications regarding a fire require skilled personnel to manage a complex of dedicated and general-use radio frequencies. Different radio bands may be used for internal communications among crew members, between the supervisor and overhead teams, and between ground and aerial operations. Special care is needed that radio frequencies be compatible, especially if resources such as fire engines and airplanes are brought in from other localities. Trained technicians ensure that communications networks are set up and do not conflict with local infrastructure.

Effective communications between supervisors and among subordinates is especially important, because of the paramilitary organization structure and because crewmembers rely upon each other for safety and support. Liaison positions are also important with the increasing use of non-English speakers on the firelines.

Specialized positions (such as fire information officer) are created within the fire organization to provide news media updates and to communicate with interested publics. Other positions may be established to interact with local communities in contending with emergency evacuations, road closures, the trauma of fatalities, and personal/property losses. Individuals receive specialized training to hold these positions and must have strong oral and written communication skills, as well as general backgrounds in fire management, fire ecology, and related subject areas.

Safety

Wildland firefighters are indoctrinated always to consider safety first when combating a wildfire. Loss of life in fighting a wildfire is simply unacceptable, no matter how valuable the resources at risk. Rookie recruits receive early training to heed the ten standard firefighting orders and the eighteen "watch out" situations while working on the fireline. Even seasoned veterans can benefit from review and reinforcement of the principles underlying the standards, especially when confronted with fires of increasing complexity and risk.

The ten standard orders were developed in 1957 by a task force established to prevent firefighter injuries and fatalities, following the Inaja fire in southern California in which eleven firefighters were killed. The eighteen situations were identified shortly afterward, and since then the orders have become non-negotiable requirements for conducting safe firefighting operations, linked to the rights of fire crews to safe working conditions. Furthermore, strict adherence to the orders should ensure public safety as well.

Ten Standard Fire Orders required of wildland firefighters are listed below (National Wildfire Coordinating Group 1993).

1. Fight fire aggressively but provide for safety first.
2. Initiate all action based on current and expected fire behavior.
3. Recognize current weather conditions and obtain forecasts.
4. Ensure that instructions are given and understood.
5. Obtain current information on fire status.
6. Remain in communication with crewmembers, your supervisor, and adjoining forces.
7. Determine safety zones and escape routes.
8. Establish lookouts in potentially hazardous situations.
9. Retain control at all times.
10. Stay alert, keep calm, think clearly, and act decisively.

The ten orders are inter-related and reinforce one another. For example, orders 2, 3, and 5 relate to fire behavior. Orders 4, 6, and 9 relate to organizational control, in terms of understanding orders and communicating effectively. Order 9 also could apply to controlling one's emotions or deflecting panic. Standards 7, 8, and 10 identify key safety requirements, while the first order in some respects provides an overarching mandate to firefighters. Gleason (1991) condensed some of the above concerns into four basic safety elements key to safe procedures for firefighters. His LCES system (Lookouts, Communications, Escape Routes, and Safety Zones) was developed to reinforce the essential elements of the standard fire orders.

A safety zone is a location where threatened firefighters can find adequate refuge from heat and rolling debris. Firefighters are sometimes encouraged to "get to the black" if confronted by an approaching fire front, meaning that a good safety zone may be provided by a burned area with minimal reburnable vegetation. As a general guideline, safety zones should be large enough to accommodate all threatened firefighters with a minimum distance of at least four times the maximum approaching flame length (www.nifc.gov/sixminutes). The LCES system places emphasis on firefighters having multiple routes for reaching a safety zone, in order to provide an extra measure of protection in case a fire blows up and cuts off access to a single route. Good communications are stressed, in terms of message clarity between lookouts and crew members and understanding of safety zone locations. Furthermore, subordinates are empowered to pose sensible questions about expectations and orders from higher authorities in the incident organization.

Specific situations in which the standard orders should be especially heeded are noted in the eighteen "watch out" situations (National Wildfire Coordinating Group 1993). These were developed after the fire orders were conceived in order to provide specific instances in which firefighters might encounter danger:

1. The fire is not scouted and sized up.
2. You are in country not seen in daylight.
3. Safety zones and escape routes are not identified.
4. You are unfamiliar with weather and local factors influencing fire behavior.
5. You are not informed of tactics, strategy, and hazards.
6. Instructions and assignments are not clear.
7. No communications link has been established with crewmembers or your supervisor.
8. You are constructing a fireline without a safe anchor point.

9. You are building fireline with fire below.
10. You are attempting a frontal assault on the fire.
11. There is unburned fuel between you and the fire.
12. You cannot see the main fire and are not in contact with someone who can.
13. You are on a hillside where rolling material can ignite fuel below you.
14. The weather is becoming hotter and drier.
15. The wind is increasing or changing direction.
16. You are getting frequent spot fires across the line.
17. The terrain and fuels make escape to safety zones difficult.
18. You are taking a nap near the fireline.

Over time, the ten standard orders and eighteen "watch out" situations have become widely acknowledged at all times for all incidents. The orders are firm—not to be broken or relaxed under any circumstances. In fact, the ten orders are invoked when investigators seek to assign causal factors for fatalities or accidents on fires. The eighteen "watch out" situations point out circumstances historically associated with firefighter fatalities and injuries. Surprisingly, many mishaps occur on smaller fires or on isolated portions of larger fires (Dear 1995). Tragedy fires commonly arise on fires that are innocent in appearance prior to flare-ups or blowups, even on the mop-up stages of some fires. Other common denominators previously identified include flare-ups in deceptively light fuels, fire runs on steep slopes with chimney canyons, and fires fanned by helicopter rotor washes or blasts of air associated with air tankers (Wilson 1977). Sadly, deaths and injuries (as well as near-misses) recur with disturbing regularity despite documentation of the dangers and the taking of necessary precautions.

Personal Protective Equipment (PPE)

As a precautionary measure, government firefighters commonly wear shirts and pants made of Nomex, the brand name for a high-strength, fire-resistant synthetic fabric generically known as aramid. Standard firefighting clothing includes hard-hat, gloves, and sturdy work-boots.

Every firefighter carries a personal fire shelter, made of aluminum foil glued to fiberglass cloth (Fuller 1991) fashioned into a pup-tent. Designed as a last resort to be deployed when all else fails, fire shelters have endured considerable controversy since their inception in 1960—including claims that they provide inadequate protection. Given sufficient warning, firefighters are expected to clear a 4 x 8 foot (1.2 x 2.4 m) area, remove the shelter from its packaging, and then deploy and climb into the tent in order to escape the heat and flames of an oncoming fire. The firefighter is instructed to wear gloves and, while lying prone, use arms and legs to hold the tent's edges down in the face of oncoming winds. Shelters are intended to protect the lungs and air passages by providing a cooler air pocket within the tent, although firefighters will sometimes deploy shelters to deflect heat while attempting to escape a fire. Multiple shelters draped end-to-end and plastered together are also sometimes used to protect historic cabins or buildings in the backcountry from oncoming fires.

Shelters are deployed in emergencies only and do not immunize a firefighter from injury. In fact, some firefighters have perished inside their shelters, resulting in calls for better materials. Following the South Canyon fire (Chapter 1), many calls were issued for improvements in shelter fabric and design; these improvements were eventually implemented in 2003. In the United States, standards and technical specifications for clothing and equipment are maintained through the USDA Forest Service Missoula Equipment Development center (http://www.fs.fed.us/eng/techdev/mtdc.htm). Tests of various materials are conducted in laboratory and field burning trials designed to test conformance to heat stress loads, portability, toxicity, and other standards.

Fire Rehabilitation

After a fire is declared out, attention switches to flood and erosion control, land rehabilitation, and other activities aimed at remediation. The intention is to prevent another disaster—for example, flooding and overland flow—following the wildfire, but also to allow the land to heal from both natural and human disturbances. The main culprit might be soil erosion from denuded hillsides, but constructed firelines by hand crews and bulldozers also may require rehabilitation. Soils can become hydrophobic or water-repelling (see Postfire Effects on Soils/Hydrology, pp. 174–176).

As a consequence, interdisciplinary teams, including soil and rangeland scientists, hydrologists, geologists, botanists, foresters, and archeologists, often are dispatched to large fires to assess the effects and develop immediate plans for rehabilitating burned areas. So-called Burned Area Emergency Rehabilitation (BAER) teams evaluate burn severity maps and develop priorities for postfire treatments. The process begins with an assessment of fire severity, using aerial imagery to classify damage to native plants, streams, wildlife habitat, archeological sites, and watersheds. Strategies include mulching, aerial seeding, tree planting, and felling dead snags or placing straw bales or wattles along the slope contour or in gullies to stabilize soil movement. Increases in ground cover seem especially important in stemming postfire erosion, so mulching or hydro-mulch sprays can be an effective tool for reducing overland soil movement (MacDonald 2004). Firelines constructed by hand crews or bulldozers also may be rehabilitated by installing water bars and trenching to slow down or redirect water flows. Hydrophobic soils may be broken up with raking or mulching charcoal into the soil to soak up excess water.

The effectiveness of some of these activities can be quite controversial, as in the case of the emergency seeding of watersheds. BAER team activities in particular can be controversial, especially if exotic weed species are introduced via seed mixes or vehicle traffic into a burned area. Establishment of exotic grasses may inhibit recovery of native plant cover, especially if the invasive plants are more effective in competing for water and nutrients. Cheatgrass (*Bromus tectorum*), an exotic annual weed, is especially problematic throughout the Great Basin region, having invaded open rangelands and fostering an insidious weed-wildfire cycle in some areas.

Fire Salvage

Another aftermath of a wildfire may be the thousands of dead and dying trees left on hillsides across the landscape. Heat from a fire can cause crown scorch/consumption or death to sensitive growing tissues (bark cambium, branches, or roots), resulting in death and injury to residual trees. Those trees not killed directly by heat may still fall prey to insects and diseases. The prospects of using fire-killed timber may seem obvious for a wood-products industry ever on the lookout for a steady supply of timber. On the other hand, environmentalists may view salvage efforts as yet another assault on the land by greedy loggers and timber companies. The salvage of fire-killed trees from public lands has thus emerged as yet another battle ground pitting commercial versus environmental interests.

The rationale behind salvage efforts after a wildfire are similar in many respects to the arguments in favor of removal and sanitation of trees killed by insect attacks, such as the mountain pine beetle (*Dendroctonous spp.*), or any other disturbance. For example, infrequent windstorms or mountain tornadoes can result in widespread tree kill in the West, just as hurricanes level vast forested areas along the East coast. In all such cases, the dead and dying trees constitute a fire hazard with a relatively short window for removal and transport to mills for processing. Besides, most people don't like to see a forest of dead and dying trees, whether killed by wildfires, insects, or any other natural disturbance.

In reality, wildfires can leave behind considerable amounts of dead, flammable biomass that may need removal. Future wildfire potential should be of acute interest whenever large amounts of woody material are allowed to accumulate over a widespread area, but other factors need to be evaluated as well. Historical fire regime, land management objectives, sustainability of excessive fuel loads, accessibility, market value, and spatial distribution of fuel hazards across the landscape all need to be considered, at minimum. In particular, understanding of the historical fire regime will provide insight into potential ignition sources, frequencies, and anticipated severity/effects of future fires. In remote and inaccessible areas the concentration of excess fuel may be of little consequence. However, in or near the urban interface, fuel accumulations should probably be reduced to lessen the severity of future fires.

Fire Restoration

The ecological restoration of fire is a growing area of interest, especially when perceived as a plausible solution to the catastrophic wildfires of the late twentieth and early twenty-first centuries. Accordingly, scientists suggest that ecosystems can and should be restored structurally to resemble forests and wildlands as they existed prior to European settlement in North America— that is, with sustainable levels of surface fuels, reduced ladder fuels, few large trees, and relatively open canopies. A forest thus restored should be less prone to devastating crown fires but also able to withstand periodic surface fires and perhaps even prosper. The rationale behind this goal relies on a combination of scientific studies (mostly the dating of historical fire-scars using tree rings and temporally paired photographs) and informed speculation about the structure (size distribution and species composition) of forests prior to European settlement. Scientific studies indicate the unequivocal disparity between the low number of contemporary ignitions versus the historical higher frequencies of lightning and native firings that originally characterized the area but that ceased with the institution of fire exclusion policies. The journals of early explorers, news articles, and oral histories all reinforce the findings from photo comparisons, suggesting that many of today's forests—in comparison with yesteryear's open, parklike stands—are choked with dense thickets of small-diameter, shade-tolerant trees. Devastating wildfire episodes only reinforce the comparisons and emphasize the need for forest restoration.

Differences in fire severity can profoundly affect the heat pulses, fuel consumption, and subsequent fire effects in an area that has undergone restoration. For example, the high-severity 2002 Hayman fire in Colorado burned as a crown fire through ponderosa pine (*Pinus ponderosa*) stands with excessive surface fuels and overgrown understory resulting in part from fire exclusion; the fire had an energy equivalent of 8 liters (roughly 2 gallons) of gasoline released per square meter burned. By contrast, stands with fuels reduced by mechanical thinning and prescribed fire might release a sixteenth as much energy (personal communication, Dr. Dan Binkley, professor of forest soils, Colorado State University).

Areas requiring ecological restoration may be most apparent in lower-elevation, long-needled pine forests (such as ponderosa pine), where photo evidence and fire scar analyses document the changes in forest structure resulting from twentieth-century fire exclusion. These areas have received the highest priority for treatment by government plans for restoring the health of public forests. Restoration efforts may not be needed in higher-elevation forests where stands dominated by species such as lodgepole pine (*Pinus contorta*) typically burn as severe, stand-replacing crown fires every few centuries. Shrublands and boreal forests also burn with characteristic high severity. Such forests may not have changed as much during the same period. Similarly, coastal Douglas fir (*Psuedotsuga menziesii*) forests in the Pacific Northwest probably have not developed a fuels accumulation problem on account of fire exclusion, since fires were relatively infrequent historically, despite large amounts of biomass. Fires in cool, moist forests may have been more easily suppressed while small in size, but the ecological effects of fire exclusion are less well studied in these higher-elevation or colder and wetter zones.

The typical tools for restoring ecosystem biodiversity include mechanical thinning and prescribed fires, the same as for fuel treatments to reduce wildfire hazards. Lightning ignitions also may be allowed to burn with minimal intervention in higher-elevation forests. However, restoration objectives imply a much broader intent to rehabilitate or mimic the ecological functioning of forests, including sustaining or restoring wildlife and insect populations and soil and water resources, as well as perpetuating endangered species—all of which may have been harmed or disrupted by human management activities. These objectives may be more complex and difficult to implement than hazard reduction alone.

Although scientifically and emotionally appealing, the idea of forest health restoration is not without its detractors. In addition to those suspicious of increased logging activity on national forests, some dispute whether the pre-European forest was truly healthier or more desirable. Others correctly point out that large fires occurred in prehistoric times, and that picking a specific pre-1900 target year for emulation of forest conditions is problematic—if not nonsensical. The desire to break the seemingly endless cycle of wildfire devastation and human tragedy seems to be the only point of general consensus.

Large-scale restoration experiments and projects have been implemented in several areas of the western United States. These usually involve collaborative partnerships that reach consensus on thorny issues such as commercial logging, or diameter limits on thinned trees. Harvesting activities are likely to increase as a result of the Healthy Forests Restoration Act of 2003. Monitoring, evaluation, and adaptive management strategies will be crucial for assessing the success of these expanded programs.

Fire Costs

Firefighting can be expensive, with recent federal expenditures routinely exceeding $1 billion per year. Expenditures generally fall into two categories: budgeted and emergency. Budgeted expenses cover preparedness (for example, personnel and equipment), but agencies tap into a separate emergency fund once a wildfire breaks out. The U.S. Forest Service budget for fire management preparedness in 2004 is around $1.7 billion, while Department of Interior agencies spend $750 million in appropriated funds. Emergency expenditures on wildfires may be much higher and more difficult to track. Previous studies (for example, USDA Forest Service 1995) illustrate the difficulties with tracking fire costs for large fire incidents, especially when multiple agencies are involved over a prolonged campaign—although record-keeping seems to have improved in recent years along with the need for increased accountability. Generally, costs increase as an incident grows in complexity and organization. A simple fire, attacked with local suppression resources, will be relatively inexpensive compared with an extended attack fire. The next level of complexity might involve a Type II incident overhead team supervising crews and resources called in from a variety of organizations at the state or regional level. If the Type II team is unable to control the fire, a national Type I overhead team may be requested, consisting of around twenty-five to thirty individuals, each with specialized assignments. The Type I team may provide oversight to thousands of firefighters in the trenches and many support activities in fire camp. The high cost of firefighting includes labor and expensive equipment used by the overhead teams, but also support services required to transport, clothe, feed, and process firefighter activities.

Individual firefighter costs are small in comparison with other resources that may be called upon during a fire, such as aircraft, but they add up quickly with the hundreds and thousands necessary to fight a fire. Representative costs for outfitting an individual firefighter may run to about $500, exclusive of boots (which are a personal responsibility). Expenses include those listed in Table 5–7, which are based on estimates provided at www.nifc.gov.

Aircraft are usually the most expensive equipment used to fight fires. In the United States, the federal government contracts with private companies to provide helicopters and airplanes. Exclusive-use contracts ensure that aircraft are available when needed but also carry costs for remaining available (that is, on standby) while waiting a fire call. Additional costs per flight hour pertain once aircraft become airborne en route to and from a fire.

Table 5-7
Cost to Outfit one Firefighter

Item	Unit cost
Fire Resistant Pants	$59.37
Fire Resistant Shirt	$45.22
Helmet with Chin Strap	$13.54
Goggles	$10.81
Gloves	$11.93
Leather Boots	(personal cost)
Head Light with Batteries	$24.02
Rations (Meals Ready to Eat-MRE)	$6.76 (each)
Canteens with Cover (4)	$11.00
Web Gear	$19.56
Web Gear Belt	$11.03
Personal Gear Bag	$27.77
Red Pack	$45.18
Fire Shelter	$40.76
Sleeping Bag	$33.88
Pulaski (hand tool)	$48.55
Shovel (hand tool)	$20.57
McLeod (hand tool)	$48.68

Source: Gonzalez-Caban 2002.

Table 5-8
Representative Airtanker Costs

Airtanker (slurry capacity, gal)	Daily Standby Cost ($ US)	Flight cost ($ US/flight hour)
P3 Orion (3,000)	$1,333	$3,250–$3,262
DC-6 (2,450)	$2,979	$2,597
PB4Y (2,200)	$1,575	$2,044–$2,276
C-130 (3,000)	$1,476	$4,024
Air Tractor 802 (800)	$1,700	$1,300
DC-4 (2,000)	$2,583	$1,884
Lead plane		$380
Slurry costs $1.00 per gallon		

Source: Gonzalez-Caban 2002

The costs listed in Table 5–8 are indicative of large air tanker capacity and lead plane costs, prior to the cancellation of the contract in May 2004. Cancellation of the contract will lead to increased reliance on smaller-capacity, lower-cost delivery planes.

Like air tanker costs, helicopter flight costs can also be quite high on an hourly basis (Gonzalez-Caban 2002). The termination of air tanker contracts did not affect helicopter services. Representative helicopter costs (1 bid.) include:

Puma Helicopter: $11,000 (per flight hour); $3,000 for support.

Sky Crane: $29,500 (per flight hour); $7,000 for support.

Lama: $1,300 (per flight hour).

Bell 206: $900 (per flight hour).

Crew costs depend on base pay (that is, for an eight-hour day), overtime, and hazardous duty differentials. On federal fires, several different pay scales may be in effect: general service (GS), administratively determined (AD), and special rates (for example, for prison inmates). Only those on GS pay scales may be paid overtime (time and a half) and hazardous duty differentials (25 percent) on top of normal base pay. In addition, government employees are entitled to lodging and meal reimbursement per diem (approximately $154/day in 2002). Typical wildfire crew costs are shown in Table 5–9.

The above sample expenses provide at best a partial estimate for the range of suppression costs that might be incurred on an incident, depending on the resources deployed to the fire. In addition, a variety of support services may be mobilized, depending on the incident size and complexity. Examples include the following:

Transportation: for example, crew transport, maintenance, water tenders, ambulances, and bulldozers;

Telecommunications, including radios and wireless networked computers;

Food, including catering units and kitchen staffs;

Waste management and sanitation;

Table 5-9
Typical Wildfire Crew Costs

Crew type	Cost
Type I Incident Management Team (35 persons)	$10,000
Hotshot (20 persons)	$6,000
Type II Crew	$3,500
Helitack (7 persons)	$1,700
Fire Engine (3 persons)	$1,000

Source: Gonzalez-Caban 2002.

Emergency medical facilities and personnel;

Reconnaissance, monitoring and surveillance, such as aerial remote sensing platforms with global positioning capabilities and remote automated weather services;

Miscellaneous supplies, such as order forms, clothing, and tools.

In comparison with other societies, U.S. fire costs are high. For example, Russia has the world's first and largest aerial firefighting force (with more than 4,000 smokejumpers and helirapellers, down from the 8,000 employed by the Soviet Union), spread across eleven time zones with an annual budget of $32 million—equivalent to a few days' expenditures during a heavy U.S. siege (Hodges 2002). The Russian aerial firefighting organization, Avialesookhrana, apparently cuts costs by using low-tech tools and keeping payrolls low (for example, $100/mo per smokejumper in 2002), and doing without fire shelters, Nomex, or seat belts on their airships. Like their U.S. counterparts, Russian smokejumpers jump out of airplanes as much as anything for the thrill.

Optimizing Fire Management Expenditures

All of the above suppression activities come at a price that must be accounted for in evaluating the cost of firefighting. Escalating suppression costs have become a major issue for public agencies with fire management responsibilities and for the taxpayers who foot the bill. Even so, managing fire damages and benefits may be as significant as suppression costs, from a policy or ecological standpoint.

Costs such as those noted in the above tables are important in ascertaining how much is spent in fighting fires, but in reality the economic efficiency of aggressive suppression versus other fire management alternatives is largely unknown. Many schemes have been developed for assessing economic efficiency over time, such as minimum cost plus loss (Sparhawk 1925) or minimum cost plus net value change (Mills and Bratten 1982), but they may only approximate the true costs and impacts to society. To show some of the theory involved with balancing costs and returns from fire management, let's look at a simplified example, using the hypothetical area illustrated in Figure 2–1 on page 54.

Recall that the example area in Figure 2–1 was zoned into seven management units, comprising two urban interface units, four commercial forest units, and one wilderness unit. Furthermore, suppose that the current management situation calls for aggressive suppression in the urban interface, fire restoration and multiple uses in the commercial forest units, and fire use in the wilderness unit—with current fire management expenditures of $10 million and annual losses (net of any fire benefits) of $20 million, or a total cost plus adjusted loss of $30 million.

Now suppose that in an upcoming election, voters will be presented with three additional alternatives to the current situation. Alternative 1 calls for building and equipping another fire station outside each financial center (at an additional cost of $2.5 million each), thereby increasing initial attack capabilities to reduce projected adjusted losses to $13 million from the current $20 million. Alternative 2 calls for building and equipping just one additional fire station ($2.5 million), but investing an additional $1.5 million in mechanical thinning and prescribed fire projects in the commercial forest units. The projected adjusted loss from Alternative 2 is $11 million. Alternative 3 will build no fire stations but invest only the additional $1.5 million in fuels treatment, with a projected adjusted loss of $15.5 million.

Hypothetical comparison of cost and losses (adjusted by benefits) for the current situation and three proposed alternatives for the hypothetical fire management area in Figure 2–1, in $millions, might resemble the listing in Table 5–10.

According to this simplified example, Alternative 2 is the best choice for the electorate—that is, building and equipping one

Table 5-10
Fire Management Costs and Losses for a Hypothetical Fire Management Area

	Current	Alternative 1 (Two additional fire stations)	Alternative 2 (additional fire station, increased fuels management)	Alternative 3 (Fuels Management only)
Costs	10.0	15.0	14.0	11.5
Adjusted loss	20.0	13.0	11.0	15.5
Total	30.0	28.0	25.0	27.0

additional fire station and investing in mechanical thinning and prescribed fire projects, at a total cost plus adjusted loss of $25 million (minimum of the alternatives considered, also known as the most efficient level, or MEL). The other alternatives reduce total costs and losses from the current situation, but not as much as Alternative 2.

Although the theory may seem simple enough from this hypothetical example, several difficulties hinder application of this approach to finding optimal expenditure levels for fire management. First, determining all costs may seem straightforward but can be complicated if multiple agencies provide fire management services in an area (that is, federal, state, local, and volunteers). Second, a determination of actual losses can be difficult to specify in advance. No one really can predict with certainty the extent to which investments in additional fire suppression forces or fuel treatments (or both) will reduce either overall expenditures or losses from future fires, subject to all the vagaries of fire weather, fire behavior and effects, and uncertain fire outbreak probabilities. Economists refer to this problem as one of determining the production function linking costs to returns, and acknowledge that its determination can be a thorny issue for fire analysts. Lastly, the costs and adjusted loss approach may not adequately reflect the important ecological, sociopolitical, and technological concerns—for example, with sustaining natural ecosystems or dealing with people who may resent being taxed for fire protection in the first place.

In spite of these problems, the analytical approach illustrated above is useful in terms of illustrating the fact that economic principles can be applied to fire management problems, although only with approximations and perhaps as only one among several criteria for making important decisions. Ecology, politics, and social concerns also need to be considered. Ultimately, a collaborative fire planning process that involves local interest groups in developing and implementing strategies across multiple jurisdictions will likely enjoy the greatest success in managing fires.

Employment Opportunities in Fire Science

To conclude this chapter on facts and data, we will look at employment opportunities in fire science and management, starting with basic positions. High school or college students interested in

future study or employment possibilities might review this section with an eye toward future career paths, although all individuals need to follow their own aspirations according to their own
personal circumstances. For many, seeing fire on the ground, or
from the end of a shovel, provides an important and invaluable
starting point—although rewarding careers are possible from
other starting points.

Firefighter

Most firefighting jobs are seasonal, being restricted to parts of the
year when temperature and moisture levels permit fires to burn.
Typically, U.S. government and state agencies recruit seasonal
employees to fill temporary jobs fighting fires across the country,
but especially in the Western states. Private contractors may provide heavy equipment, crew personnel, and support services
such as meals, showers, and portable toilets in fire camps. Many
temporary employees are college students and teachers, who join
the workforce during summer break. Native tribal organizations
may organize suppression teams that are available for dispatch to
fires or as fuels thinning crews. Some crews are made up of migrant laborers for whom English is a second language. For some,
seasonal employment represents an opportunity to establish a
performance record that could become a stepping stone for future
permanent employment; for others, the summer fire season may
represent the only opportunity for earning decent wages
throughout the year. Pay for temporary firefighters ranges from
$8.64 an hour to $11.64 an hour (based on rates in effect in 2002).

A typical day for seasonal firefighters might include organized physical training and project work. Even during a hot summer, firefighters will encounter periods during a forty-hour work
week when fires are not burning. During such times, they will be
employed thinning forest stands or performing other hazard reduction or fire prevention activities. However, once fires break
out, government firefighters will typically work long shifts and
receive overtime and hazardous duty pay differentials. Non-
government firefighters may not receive the overtime and hazardous duty pay.

Despite the long hours and arduous labor, many firefighters
love the work and look forward to repeating seasonal work year
after year, often at higher pay or in pursuit of a permanent work
appointment. After gaining experience on a fire crew (engine or

hand-crew), seasonal employees can apply for more elite positions such as hot-shot crews or smokejumpers. Hot-shot crews typically draw tough assignments and, like other crew organizations, develop strong camaraderie over a fire season. Smokejumpers typically attack fires in remote locations by parachuting out of airplanes and making use of firefighting equipment included in a separate drop. Helitack crews rely on helicopters for transport to remote locations. Crews often travel to other parts of the country to support firefighting efforts in other regions. An applicant for federal appointment must be at least eighteen years old, a U.S. citizen, in good physical condition, and able to pass a physical exam (including drug test).

Full-time agency employees may be called away from their normal duties (for example, as foresters, resource specialists, or biologists) to assist with firefighting assignments, especially on local incidents, but also as members of overhead teams available nationwide. Participation on fire assignments is usually an expectation of full-time agency employees. Individuals with primary fire responsibilities may qualify eventually for early firefighter retirement benefits, including full pension by age fifty with thirty years' experience.

Fire Manager or Planner

Individuals with extended seasonal firefighting experience may eventually land a full-time permanent position that focuses on fire, either year-round or mixed with collateral responsibilities in a related natural resource specialty. Fire management officer (FMO) duties include planning, preparation, and implementation of management plans for an area or jurisdiction, including overall supervision of crews and suppression resources on local fires. Of late, the FMO has been considered a professional position in terms of civil service ratings—and in terms of job expectations. Full-time assistant fire management officers (AFMO) traditionally are hired out of the technician series. The professional-technician distinction between employees is not hard and fast, but professionals generally possess college educations or advanced schooling. Typically, advancement through the career ranks depends on both on-the-ground experience and levels of college education. Exceptions occur, as in any large bureaucratic organization, but as in most fields, a college education generally improves starting pay and advancement opportunities.

Fuels specialists plan and participate on prescribed fire and other fuel treatment projects, such as mechanical thinning. Fire ecologists keep current with the latest developments in understanding fire effects and the role of fire in ecosystem processes in order to contribute to the formulation of management plans or environmental assessments. Most fire specialists participate on overhead teams and may provide input and feedback in the development of agency environmental assessments and management plans.

Fire prevention technicians perform routine inspections and patrols through the forest, with the aim of enforcing compliance with regulations and reducing ignition sources or risky human behavior. Some prevention specialists may be trained fire source investigators. Engine and crew superintendents generally work their way up the ranks from seasonal or full-time technician jobs.

Starting professional federal pay scales in 2003 ranged from $23,442 to $61,251 (GS 5–13), with step increases based on time in grade. Federal technician salaries range from $26,130 to $51,508 (GS6–12), also with step increases. Both professionals and technicians are eligible for health and retirement benefits. Federal firefighters qualify for early-retirement privileges based on age and years in service. Pay scales for state, private, and urban firefighters with wildland responsibilities may vary considerably from their federal counterparts, depending in part on union and cost-of-living differentials.

Fire Researcher

Full-time research positions involve the study of fire behavior, fire effects, and fire management, along with a diverse variety of other disciplines such as mathematics, engineering, or sociology. Employees may have job titles such as research forester or research scientist. Some positions stress basic science findings, while others focus on research, development, and applications. Technology transfer specialists apply the latest scientific findings to field applications. Government scientists working at a research laboratory or in the field usually hold a master's degree, although many begin as part-time employees while working their way through college. A Ph.D. is required for research project leaders. Fire experience, though desirable, is not required.

Starting federal pay scales in 2003 for fire researchers ranged from $29,037 to $85,140 (GS7–15), with step increases based on time in grade. Other fire research positions are found at universities and colleges, as well as in private industry. Salaries will depend on market forces and institutional policies.

Summary

People with inquisitive minds and interests in studying fire may find countless employment opportunities with the government (as managers or researchers), on academic faculties (conducting instruction and research) at universities and colleges, and in the private sector as consultants or with nongovernmental organizations. A partial listing of the employment areas may be inferred from the abbreviated list of issue or problem areas related to fires and their management. Items in the list may be related to one another or overlap, but at least they may be suggestive of the many opportunities:

- Fire management, prevention, dispatch, detection, suppression, use, rehabilitation, and equipment development and testing;
- Fire ecology, fire history, the role of fire in sustaining ecosystems;
- Fuel dynamics, assessment, modeling, and description;
- Fuel moisture, assessment, and analysis;
- Fire physics, understanding of basic mechanisms of fire spread and effects;
- Fire behavior prediction, measurement, and evaluation;
- Fuel consumption;
- Smoke management and dispersion;
- Remote sensing, cartography, and geographic information systems related to fires;
- First order fire effects, aboveground, belowground, on atmospheric and water resources, from cultural and social perspectives;
- Longer term fire effects on landscapes;
- Fire meteorology, climatology, lightning analysis, assessment, and forecasting;
- Prescribed fire implementation and effects;
- Economics, policy, and planning;
- Ecosystem engineering and forest health management;
- Wildland urban interface, county planning, incident management;

- Computer programming;
- Educational programming, written and visual media, development and distribution;
- Journalism and creative writing;
- Emergency management.

Other employment possibilities exist in disciplines related to the above list, especially for individuals with a fire science or management background. Furthermore, the age structure and demographics of the current workforce in fire and emergency management suggest wholesale turnovers in personnel over the next five years, as senior-level employees (ages fifty to fifty-five) retire or leave. A vast storehouse of experience and knowledge about fires and their management will exit with these senior employees, especially regarding the megafires—their behavior, their root causes, and the futility of certain approaches to their management. As vacancies are filled behind the upcoming departures, fire workers will become more youthful, more intellectually diverse, and more representative of the multiple cultures within our society.

Employment in the fire community, whether in management, research, academia, or some related area, can provide excitement, intellectual stimulation, opportunities to work with outstanding individuals and institutions, and much more. Fires capture the human imagination like few other natural phenomena, but they also can instill fear and frustration under certain circumstances. Often these fears and frustrations may seem to outweigh the benefits, although this response may be as much cultural perception as reality. Through education, perhaps future generations will enjoy greater success in learning to live with fires and their consequences.

Literature Cited

Agee, J. K. 1993. Fire Ecology of Pacific Northwest Forests. Washington, DC: Island Press, 493 p.

Agee, J. K., B. Bahro, M. A. Finney, P. N. Omi, D. B. Sapsis, C. N. Skinner, J. W. van Wagtendonk, and C. P. Weatherspoon. 2000. "The Use of Shaded Fuelbreaks in Landscape Fire Management." *Forest Ecology and Management* 127: 55–66.

Agee, J. K., and C. N. Skinner. 2003. "Ecological Principles of Forest Fuel Reduction Treatments." Presentation at Risk Assessment related to Decision-making related to Uncharacteristic Wildfire, Nov. 17–19, 2003. Portland: Oregon State University, Forestry Outreach Education.

Albini, F. A. 1979. "Spot Fire Distance from Burning Trees—A Predictive Model." USDA Forest Service General Technical Report INT-56, 73 p.

———. 1983. "Potential Spotting Distances from Wind-driven Surface Fires." USDA Forest Service General Technical Report INT-309, 27 p.

Alexander, M. D. 2004. "Enhanced Crew Cohesion and a Methodology for Evaluation." M.S. professional paper, Colorado State University.

Alexander, M., C. Stefner, J. Beck, and R. Lanoville. 2001. New insights into the effectiveness of fuel reduction treatments on crown fire potential at the stand level. Page 318 *in* G. Pearce and L. Lester (technical coordinators). Bushfire 2001 Conference Proceedings, 3–6 July 2001. Christchurch: New Zealand Forest Research Institute.

Anderson, H. E. 1982. "Aids to Determining Fuel Models for Estimating Fire Behavior." USDA Forest Service General Technical Report INT-122, 22 p.

Andrews, P. L. 1986. "BEHAVE: Fire Behavior Prediction and Fuel Modeling System—BURN Subsystem, Part 1." USDA Forest Service General Technical Report INT-194, 130 p.

Andrews, P. L., and C. H. Chase. 1989. "BEHAVE: Fire Behavior Prediction and Fuel Modeling System—BURN Subsystem, Part 2." USDA Forest Service General Technical Report INT-260, 93 p.

Astaris Corporation 1998. "Phos-Chek Foam: Making Water a Better Firefighter." Astaris LLC, St. Louis. Nonpaginated promotional materials.

Bevins, C. 2003. *BehavePlus Fire Modeling System.* Missoula, MT: USDA Forest Service, Rocky Mountain Research Station and Systems for Environmental Management.

Boxall, B. 2002. "Intense Logging Blamed for Wildfires." Los Angeles *Times,* September 17, 2002, p. A8.

Bradstock, R. A., J. E. Williams, and A. M. Gill. 2002. *Flammable Australia: The Fire Regimes and Biodiversity of a Continent.* Cambridge: Cambridge University Press, 462 p.

Brown, A. A., and K. P. Davis. 1973. *Forest Fire: Control and Use.* 2d ed. New York: McGraw-Hill Book Co., 686 p.

Brown, J. K. 1974. "Handbook for Inventorying Downed Woody Material." USDA Forest Service General Technical Report INT-16, 24 p.

———. 2000. "Chapter 1: Introduction and Fire Regimes." Pp. 1–6 in *Wildland Fire in Ecosystems: Effects of Fire on Flora,* edited by J. K. Brown and J. K. Smith. USDA Forest Service General Technical Report RMRS-GTR-42, vol. 2, 257 p.

Brown, J. K., R. D. Oberheu, and C. M. Johnston. 1982. "Handbook for Inventorying Surface Fuels and Biomass in the Interior West." USDA Forest Service General Technical Report INT-129. 48 p.

Brown, J. K., and J. K. Smith, eds. 2000. "Wildland Fire in Ecosystems: Effects of Fire on Flora." USDA Forest Service General Technical Report RMRS-GTR-42, vol. 2, 257 p.

Brown, P. M., M. R. Kaufmann, and W. D. Shepperd. 1999. "Long-term Landscape Patterns of Past Fire Events in a Montane Ponderosa Pine Forest of Central Colorado." *Landscape Ecology* 14: 513–532.

Brown, P. M., and W. D. Shepperd. 2001 "Fire History and Fire Climatology along a 5∞ Gradient in Latitude in Colorado and Wyoming, USA." *Paleobotanist* 50: 133–140.

Burgan, R. E., and R. A. Hartford. 1993. "Monitoring Vegetation Greenness with Satellite Data." USDA Forest Service General Technical Report INT-297, 13 p.

Bustillo, M. 2003. "Scientists Predict Long-term Slide Risk." Santa Rosa *Press Democrat*, December 27, 2003, p. A10.

Byram, G. M. 1959. "Combustion of Forest Fuels." Pp. 61–89 in *Forest Fire Control and Use*, edited by K. P. Davis. New York: McGraw-Hill Book Co., 584 p.

Campbell, D. 1991. *The Campbell Prediction System*. Ojai, CA: Ojai Printing and Publishing Co.

Chandler, C., P. Cheney, P. Thomas, L. Trabaud, and D. Williams. 1983a. *Fire in Forestry*. Vol. I: *Forest Fire Behavior and Effects*. New York: John Wiley and Sons, 450 p.

———. 1983b. *Forestry*. Vol. II: *Fire Management*. New York: John Wiley and Sons, 298 p.

Countryman, C. n.d. "Fuel Moisture and Fire Danger—Some Elementary Concepts." USDA Forest Service Region 5, Aviation and Fire Management handout, 4p.

———. 1966. "The Concept of Fire Environment." USDA Forest Service, *Fire Control Notes* 27(4): 8–10.

———. 1972. "The Fire Environment Concept." USDA Forest Service Research Paper. Berkeley: Pacific Southwest Research Station, 12 p.

Dear, J. A. 1995. Letter to Jack Ward Thomas, chief, USDA Forest Service. February 8, 1995, 2p plus attachment.

DeBano, L. F. 1981. "Water Repellent Soils: A State of the Art." USDA Forest Service General Technical Report PSW-GTRR-46, 21 p.

DeBano, L. F., D. G. Neary, and P. F. Folliott. 1998. *Fire Effects on Ecosystems*. New York: John Wiley and Sons, 333 p.

Deeming, J. D. 1988. "Effects of Fire on Air Quality." Lesson plan, introduction to Fire Effects course, Washington Institute, Redding CA.

———. 1990. "Effects of Prescribed Fire on Wildfire Occurrence and Severity." Pp. 95–104 in *Natural and Prescribed Fire in Pacific Northwest Forests,* edited by J. D. Walstad, S. R. Radosevich, and D. V. Sandberg. Corvallis: Oregon State University Press, 317 p.

Deeming, J. E., R. E. Burgan, and J. D. Cohen. 1977. "The National Fire-Danger Rating System, 1978." USDA Forest Service General Technical Report INT-39, 63 p.

DeHaan, J. D. 2002. *Kirk's Fire Investigation.* 5th ed. Upper Saddle River, NJ: Prentice-Hall. 638 p.

Dennis, F. C. 1992. "Creating Fire Safe Zones around Your Forested Homesite." Colorado State University Extension, Service in Action sheet no. 6.302, 4 p.

Doolittle, M. L., and M. L. Lightsey. 1980. "Southern Woods-burners: A Descriptive Analysis." USDA Forest Service Research Paper S0–151, 6 p.

Enders, J. 2001. "Firefighting Effort Boosted by Helicopters." Ft. Collins *Coloradoan,* August 18, 2001, 2.

Finney, M. A. 1998. "FARSITE: Fire Area Simulator." USDA Forest Service Research Paper RMRS-RP-4. 47 p.

———. 2001. "Design of Regular Landscape Fuel Treatment Patterns for Modifying Fire Growth and Behavior." *Forest Science* 47: 219–228.

Fischer, W. C. 2003. "Fire Management." Chapter 8 in *Introduction to Forest and Renewable Resources,* 7th ed., edited by G. W. Sharpe, C. W. Hendee, and W. F. Sharpe. Boston: McGraw Hill, 544 p.

Fuller, M. 1991. *Forest Fires: An Introduction to Wildland Fire Behavior, Management, Firefighting, and Fire Prevention.* Wiley Nature Editions. New York: John Wiley and Sons, 239 p.

Gleason, P. 1991. "LCES—A Key to Safety in the Wildland Fire Environment." *Fire Management Notes* 52(4): 9.

Gonzalez-Caban, Armando. 2002. "Costes de los medios aereos de combate de incendios forestales en el servicio forestal del departamento de Agricultura de los EEUU: Una primera aproximacion." Pp. 65–72 in *I simposio internacional La Gestion de los Medios Aereos en la Defensa Contra los Incendios Forestales,* edited by Antonio J. Gonzalez Barrios and Ignacio Ribas i Sole, coordinadores, Cordoba, Octobre 28–30, 2002. Cordoba, Espana: Universidad de Cordoba, 367 p.

Gorte, R. 2000. "Forest Fire Protection." Congressional Research Service Report for Congress Order Code RL30755, the Library of Congress, Washington, DC, 29 p.

Graham, R. T., tech. ed. 2003. "Hayman Fire Case Study." General Technical Report RMRS-GTR-114. Ogden, UT: U.S. Department of Agriculture, Forest Service, Rocky Mountain Research Station, 396 p. Also available at http://www.fs.fed.us/rm/pubs/rmrs_gtr114.html.

Green, L. R. 1977. "Fuelbreaks and Other Fuel Modification for Wildland Fire Control." USDA Ag. Handbook 499.

Haines, D. A., and R. W. Sando. 1969. "Climatic Conditions Preceding Historically Great Fires in the North Central Region." USDA Forest Service Research Paper NC-34. 19 p.

Hardy, C. C., J. P. Menakis, D. G. Long, J. K. Brown, and D. L. Bunnell. 1998. "Mapping Historic Fire Regimes for the Western United States: Integrating Remote Sensing and Biophysical Data." Pp. 288–300 in *Proceedings of the 7th Biennial Forest Service Remote Sensing Applications Conference,* April 6–9, 1988, Nassau Bay, TX. Bethesda, MD: American Society for Photogrammetry and Remote Sensing.

Heinselman, M. L. 1978. "Fire in Wilderness Ecosystems." Pp. 249–278 in *Wilderness Management,* edited by J. C. Hendee, G. H. Stankey, and R. C. Lucas. Miscellaneous publication no. 1365. Washington, DC: USDA Forest Service.

Hodges, G. 2002. "Smokejumpers." *National Geographic* 204(8):82–99.

Joesten, M. D., J. T. Netterville, and J. L. Wood. 1993. *World of Chemistry— Essentials.* Fort Worth, TX: Saunders College Publishing.

Johnson, E. A. 1992. *Fire and Vegetation Dynamics: Studies from the North American Boreal Forest.* Cambridge: Cambridge University Press, 129 p.

Keetch, J. J., and G. Byram. 1968. "A Drought Index for Forest Fire Control." USDA Forest Service Research Paper SE-38. Revised 1988. Asheville, NC: USDA Forest Service Southeastern Forest Experiment Station, 32 p.

Kelha, V., Y. Rauste, and A. Buongiorno. N.d. "Forest Fire Detection by Satellites for Fire Control." Available at http://www.vtt.fi/tte/research/tte1/tte14/proj/FF-Operat/Forest_fire_detection.pdf.

Kilgore, B. M. 1981. "Fire in Ecosystem Distribution and Structure: Western Forests and Scrublands." Pp. 58–59 in *Fire Regimes and Ecosystem Properties: Proceedings of the Conference,* edited by H. A. Mooney, T. M. Bonnicksen, N. L. Christensen, J. E. Lotan, and R. A. Reiners, tech. co-ords. December 11–15, 1978, Honolulu, HI. USDA Forest Service General Technical Report WO-26.

Kimmins, J. P. 1987. *Forest Ecology.* New York: Macmillan Publishing, 531 p.

MacDonald, L. H. 2004. "Post-fire Erosion in the Colorado Front Range." Abstract of seminar presentation February 27, 2004, at Colorado State University.

MacDonald, L. H., and J. D. Stednick. 2003. "Forests and Water: A State-of-the-Art Review for Colorado." CWRRI Completion Report no. 196. Colorado Water Resources Research Institute Colorado State University, Ft. Collins, CO. 65 p.

Martin, R. E. 1976. "Prescribed Burning for Site Preparation in the Inland Northwest." Pp. 134–156 in *Tree Planting in the Inland Northwest: Short-course Proceedings held at Washington State University*, edited by David M. Baumgartner and Raymond J. Boyd. Pullman, February 17–19, 1976, Washington State University, Cooperative Extension Service. Pullman, 311 p.

McKee, T. B., N. J. Doesken, J. Kleist, and C. J. Shrier. 2000. "What Is Drought?" P. 5 in *Water in the Balance*, Colorado Water Resources Research Institute Bulletin no. 9 (2d ed.). Colorado State University, 19 p.

Miller, M. 2000. "Fire Autoecology." Pp. 9–50 in *Wildland Fire in Ecosystems: Effects of Fire on Flora*, edited by J. K. Brown and J. K. Smith, eds. USDA Forest Service General Technical Report RMRS-GTR-42, vol. 2, 257 p.

Mills, T. J., and F. W. Bratten. 1982. "FEES: Design of a Fire Economics Evaluation System." USDA Forest Service General Technical Report PSW-65, 37 p.

Morgan, P., S. C. Bunting, A. Black, T. Merrill, and S. Barrett. 1996. "Fire Regimes in the Interior Columbia River Basin: Past and Present." Final Report for RJVA-INT-94913. On file at USDA Forest Service, Rocky Mountain Research Station, Fire Sciences Laboratory, Missoula, MT, 31 p.

———. 1998. "Past and Present Fire Regimes in the Interior Columbia River Basin." Pp. 77–82 in *Fire Management under Fire (Adapting to Change)*, edited by K. Close and R. A. Bartlette. Proceedings of the 1994 Interior West Fire Council meeting and program, November 1–4, 1994, Coeur d'Alene, ID. Fairfield WA: International Association of Wildland Fire.

Mutch, R. W. 1970. "*Wildland Fires and Ecosystems: A Hypothesis.*" *Ecology* 51: 1046–1051.

National Wildfire Coordinating Group. 1993. Look Up, Look Down, Look Around. Fireline Safety Reference NFES 2243, National Interagency Fire Center. Washington, DC: Government Printing Office, non-paginated.

Nature Conservancy. 2003. "Fire Initiative." Available at http://nature .org/initiatives/fire/#.

Omi, P. N. 1997. "Fuels Modification to Reduce Large Fire Probability." Final Report submitted to U.S. Dept. of Interior, Fire Research Committee. Western Forest Fire Research Center, Colorado State University, 51 p. plus appendices.

———. 2004. "Evaluating Tradeoffs between Wildfires and Fuel Treatments," in *Proceedings of the 2d Symposium on Fire Economics, Planning and Policy: A Global View.* April 19–22, 2004. Cordoba, Spain (CD-ROM). 10 p.

Omi, P. N., and K. D. Kalabokidis. 1991. "Fire Damage on Extensively versus Intensively Managed Forest Stands within the North Fork Fire, 1988." *Northwest Science* 65: 149–157.

Omi, P. N., and E. J. Martinson. 2002. "Effect of Fuels Treatment on Wildfire Severity." Final report submitted to the Joint Fire Science Program Governing Board. Western Forest Fire Research Center, Colorado State University, 36 p.

———. 2003. "Drought, Fire, and Fuel Treatments." In *Proceedings of the 3d International Wildland Fire Conference and Exhibition.* October 3–6, 2003. Sydney, Australia (CD-ROM).

———. 2004. "Fuel Treatments and Fire Regimes." Final report submitted to the Joint Fire Science Program Governing Board. Western Forest Fire Research Center, Colorado State University, 44 p.

Ottmar, R., D. Sandberg, C. Wright, and R. Vihanek. 2002. "Modification and Validation of Fuel Consumption Modeling." Joint Fire Science Program Progress Review, March 11–14, 2002, San Antonio, TX. CD copy of presentation.

Ottmar, R., R. Vihanek, and C. Wright. 2002. "Photo Series Phases II and III." Joint Fire Science Program Progress Review, March 11–14, 2002, San Antonio, TX. CD copy of presentation.

Palmer, W. C. 1965. "Meteorological Drought." Research Paper no. 45, U.S. Dept. of Commerce Weather Bureau, Washington, DC.

Pollet, J., and P. N. Omi. 2002. "Effect of Thinning and Prescribed Burning on Wildfire Severity in Ponderosa Pine Forests." *International Journal of Wildland Fire* 11: 1–10.

Pyne, S. J., P. L. Andrews, and R. D. Laven. 1996. *Introduction to Wildland Fire: Fire Management in the United States.* New York: John Wiley and Sons, 455 p.

Rideout, D. B., and P. N. Omi. 1995. "Estimating the Cost of Fuels Treatment." *Forest Science* 41: 664–674.

Rothermel, R. C. 1972. "A Mathematical Model for Predicting Fire Spread in Wildland Fuels." USDA Forest Service Research Paper INT-115, 40 p.

———. 1983. "How to Predict the Spread and Intensity of Forest and Range Fires." USDA Forest Service General Technical Report INT-143, 161 p.

———. 1991. "Predicting Behavior and Size of Crown Fires in the Northern Rocky Mountains." USDA Forest Service Research Paper INT-438, 46 p.

Rowe, J. S. 1983. "Concepts of Fire Effects on Plant Individuals and Species." Pp. 135–154 in *The Role of Fire in Northern Circumpolar Ecosystems*, edited by R. Wein and D. A. Maclean. London: Wiley.

Ryan, K. C. 2001. "Fire Severity: Concepts and Applications." CD-ROM version of presentation to Colorado State University, Department of Forest Sciences, May 2001.

Ryan, K. C., and N. Noste. 1985. "Evaluating Prescribed Fires." Pp. 230–237 in *Proceedings of the Symposium on Wilderness Fire*, November 15–18, 1983, Missoula, MT.

Sandberg, David V., Roger D. Ottmar, Janice L. Peterson, and John Core. 2002. "Wildland Fire in Ecosystems: Effects of Fire on Air." General Technical Report RMRS-GTR-42-vol. 5. Ogden, UT: U.S. Department of Agriculture, Forest Service, Rocky Mountain Research Station, 79 p.

Schmidt, Kirsten M., James P. Menakis, Colin C. Hardy, Wendall J. Hann, and David L. Bunnell. 2002. "Development of Coarse-scale Spatial Data for Wildland Fire and Fuel Management." General Technical Report RMRS-GTR-87. Fort Collins, CO: U.S. Department of Agriculture, Forest Service, Rocky Mountain Research Station. 41 p. plus CD.

Schroeder, M. J., and C. C. Buck. 1970. "Fire Weather." USDA Handbook 360. U.S. Department of Agriculture, Forest Service, Washington, DC, 229 p.

Smith, J. K., ed. 2000. "Wildland Fire in Ecosystems: Effects of Fire on Fauna." USDA Forest Service General Technical Report RMRS-GTR-42, vol. 1, 83 p.

Smith, J. K., and W. C. Fischer. 1997. "Fire Ecology of the Forest Habitat Types of Northern Idaho." USDA Forest Service General Technical Report INT-363, 142 p.

Sparhawk, W. N., 1925. "The Use of Liability Ratings in Planning Forest Fire Protection." *Journal of Agricultural Research* 33(8): 693–762.

Starker, T. J. 1934. "Fire Resistance in the Forest." Journal of Forestry 32: 462–467.

Stokes, M. A., and J. H. Dieterich, eds. 1980. "Proceedings of the Fire History Workshop." USDA Forest Service General Technical Report RM-81. Ft. Collins: USDA Forest Service Rocky Mountain Research Station, October 20–24, 1980. Tucson, Arizona. 142 p.

Swetnam, T. W., and J. L. Betancourt. 1990. "Fire-southern Oscillation Relations in the Southwestern United States." *Science* 249: 1017–1020.

Trollope, W. S. W. 1984. "Fire in Savanna." Pp. 151–175 in *Ecological Effects of Fire in South African Ecosystems*, edited by P. de V. Booysen and N. M. Tainton. Berlin: Springer-Verlag.

UN Economic Commission for Europe. 1982. "Forest Fires: Prevention and Control." In *Proceedings of an International Seminar Organized by the Timber Committee of the UN Economic Commission for Europe*, edited by T. van Nao Held at Warsaw, Poland, at the invitation of the government of Poland, May 20–22, 1981. The Hague: Martinus Nijhoff/Dr. W. Junk Publishers, 236 p.

USDA Forest Service. 1995. "Fire Suppression Costs on Large Fires." USDA Forest Service, Fire and Aviation Management, Washington, DC, 18 p. plus appendices.

Van de Water, R. n.d. "Wildland Fires for Resource Benefits in Wilderness Areas: The Yosemite National Park Experience." (available at http://watershed.org/news/win_00/13_yosemite.htm).

Van Wagner, C. E. 1977. "Conditions for the Start and Spread of Crown Fires." *Canadian Journal of Forest Research* 7: 23–24.

van Wagtendonk, J. W. 1995. "Large Fires in Wilderness Areas." Pp. 113–120 in *Proceedings of the Symposium on Fire in Wilderness and Park Management*, edited by J. K. Brown, R. W. Mutch, C. W. Spoon, and R. H. Wakimoto. March 30–April 1, 1993, Missoula, MT: USDA Forest Service General Technical Report INT-GTR-320, 283 p.

Wagenbrenner, J. W. 2003. "Effectiveness of Burned Area Emergency Rehabilitation Treatments, Colorado Front Range." Master's thesis, Colorado State University. 76 p.

Ward, D. E., and C. C. Hardy. 1991. "Smoke Emissions from Wildland Fires." *Environment International* 17: 117–134.

Weatherspoon, C. P., and C. N. Skinner. 1995. "An Assessment of Factors associated with Damage to Tree Crowns from the 1987 Wildfires in Northern California." *Forest Science* 41: 430–451.

Werth, J., and P. Werth. 1998. "Haines Index Climatology for the Western United States." *Fire Management Notes* 58(3): 8–17.

Whelan, R. J. 1995. *The Ecology of Fire*. Cambridge: Cambridge University Press, 346 p.

Williams, J. 2003. "Values, Tradeoffs, and Context: A Call for a Public Lands Policy Debate on the Management of Fire-dependent Ecosystems." In *Proceedings of the 3d International Wildland Fire Conference*. October 3–6, 2003. Sydney, Australia, 7 p. (CD-ROM).

Williams, V. 2003. "Evaluation of Wildfire Burn Severity Classification, Utilizing Ground and Remote Sensing Methodologies, Southern Colorado, USA." MSc dissertation, University of Edinburgh, Scotland, 99 p. (CD-ROM provided by author).

Wilson, C. C. 1977. "Fatal and Near-fatal Forest Fires: The Common Denominators." *International Fire Chief* 43(9): 9–15.

Wilson, R. 1980. "Reformulation of Forest Fire Spread Equations in SI Units." USDA Forest Service Research Note INT-292, 5 p.

Wright, H., and A. Bailey. 1982. *Fire Ecology: United States and Southern Canada.* New York: Wiley-Interscience, 501 p.

Zhang, Y., J. M. Wallace, and D. S. Battisti. 1997. "ENSO-like Interdecadal Variability: 1900–93." *Journal of Climate* 10: 1004–1020.

6

Directory of Agencies and Organizations

In this chapter we examine the myriad federal, state, and local government agencies and private organizations that play a role or have an interest in forest fire management. We focus again on the United States, but also include mention of significant organizations globally. Newcomers to the United States are often puzzled that so many different players have an active stake in fire management, or that responsibilities could vary so greatly from state to state. The reasons are rooted in the fact that states are responsible for such activities as firefighting and police activities. Despite their apparent complexity, fire management and research in the United States often set an example to be emulated elsewhere, or from which valuable lessons can be learned. In recognition of the importance of fostering international understanding, the United States has in recent years entered into reciprocal exchanges of firefighters and overhead management teams with nations such as Canada, Mexico, and Australia. As in the United States, wildfire management in other countries often involves cooperation between a complex set of land ownerships and agency jurisdictions, as for example when private land is interspersed within a national forest or park, or where city, county, state, federal, and volunteer fire organizations all coexist close to one another.

The need for interagency cooperation is accentuated by the variety of management activities that may be undertaken on a tract of land, such as fire prevention, preparedness, hazard mitigation, suppression, reclamation/rehabilitation, and public relations (see

Chapter 5)—all of which may require detailed planning and budgetary commitments. Furthermore, the set of activities will possibly require a complex set of agreements and fiscal reciprocities if multiple jurisdictions are involved. In addition, regulatory activities—such as air or water quality control—may require that agencies comply with standards or obtain permits for management activities, such as prescribed burns. Although such arrangements might seem cumbersome, over the years agencies have developed mutual aid compacts, delegated authority to one another, and generally agreed to cooperate as needed to manage the land as efficiently as possible while complying with regulations (for example, air and water quality, endangered species habitat protection). This interagency cooperation effort works reasonably well most of the time and is absolutely essential, since no agency is able to handle its fire workload independently. Still, just the paperwork necessary to facilitate cooperation and cost-sharing among agencies can be staggering—but somehow the work gets accomplished. Even so, some arrangements, such as cost reimbursement on a complex set of wildfires spreading over multiple jurisdictions, can require literally years of negotiation and compromise to settle all accounts.

Additional complications may arise where burned-over areas receive disaster designation, opening the door to victim assistance such as grants and loans, or where volunteer fire departments, metropolitan fire districts, and private contractors join the mix of federal and state firefighting resources. Also, the settling of private insurance claims for wildfire losses can be complex, especially if disputes arise over policy coverage. Some claims may end up in court and may not be resolved for years. Even before a fire, a variety of private interest groups may seek a voice in management and policy decisions. In truth, the sociopolitical environment for fire management decisions can be as complicated as the fuel and weather interactions that drive a fire across a landscape.

Fire Policy and Administration

National Wildfire Coordinating Group
Arizona State Forester
2901 W. Pinnacle Peak Road
Phoenix, AZ 85027–1002
http://www.nwcg.gov

The National Wildfire Coordinating Group (NWCG) is an interagency body that aims to coordinate and unify efforts by the numerous wildfire management agencies so as to facilitate cooperation while avoiding wasteful duplication. Member agencies include the USDA Forest Service, USDI (Bureau of Land Management, National Park Service, Bureau of Indian Affairs, and the Fish and Wildlife Service), U.S. Department of Homeland Security (U.S. Fire Administration), the Intertribal Timber Council, and state forestry agencies. Chair responsibilities rotate among member agencies. Member agencies sit on various working teams and advisory groups to provide federal guidance in different subject areas, such as the following:

- fire danger working team advances the science and application of fire danger rating;
- fire equipment working team coordinates chemical and fire equipment needs, development, and implementation among federal and state agencies;
- fire use working team coordinates and advocates the use of wildland fire to achieve resource benefits;
- fire weather working team identifies opportunities for improving fire weather services;
- incident business practices working team identifies issues, develops acceptable solutions, and implements the solutions within business practices for wildfire, nonfire, and Federal Emergency Management Agency responses among state and federal agencies;
- incident operations working team addresses issues regarding the National Wildfire Qualifications and Certifications System and the National Interagency Incident Management System.
- information resource management working team identifies policy-level information issues that may affect interagency fire management and addresses them through information and communication systems;
- wildland fire education working team provides leadership for national-level wildland fire education programs;
- publication management system working team makes all information sponsored by NWCG readily available, including publications and visual materials;

- training working team develops and coordinates fire management training programs;
- wildland urban interface working team provides leadership in addressing issues regarding fire activity and firefighter/public safety in the wildland/urban interface;
- wildland fire investigation working team develops policies, methods, procedures, and other actions for systematic and science-based wildland fire investigations to support criminal prosecution and cost recovery;
- radio narrowband advisory group identifies and addresses policy-level and operational radio narrowband issues that could affect interagency fire management activities; and
- fire social sciences task group addresses social science needs related to wildland fire management

National Interagency Fire Center
3833 S. Development Avenue
Boise, ID 83705
http://www.nifc.gov

The National Interagency Fire Center (NIFC) in Boise, Idaho, provides nationwide coordination of fire activities, as well as maintaining a cache of firefighting equipment and supplies, a smokejumper base, publications, and other activities. Affectionately called "Nif-see" by practitioners, the center implements mandates from the NWCG and its member agencies related to the management of specific programs and individual incidents. During extended wildfire sieges, NIFC sometimes becomes the national focal point for fire operations and information dissemination. Resembling a paramilitary nerve center, NIFC coordinates activities of a vast network of member agencies to supply human resources, transportation, supplies, and support services necessary for responding to wildfire incidents. Permanent NIFC members represent the same land management agencies as the NWCG, although other agencies (for example, National Association of State Foresters, U.S. Weather Bureau, and Office of Aircraft Services) are added as necessary to implement policy set by NWCG.

Wildfire management is coordinated through the National Interagency Coordination Center (NICC), which processes requests from command centers established in different geographic regions or geographic area coordination centers (GACCs). Thus requests for firefighting resources for several incidents can be considered and forwarded upward if needs cannot be met regionally by the GACC. The use of acronyms within the fire bureaucracy at first may seem overwhelming to the uninitiated, although practitioners rely on jargon and acronyms to communicate with one another.

USDA Forest Service Fire and Aviation Management
Auditor's Building
201 14th Street, SW at Independence Ave., SW
Washington, DC 20250

The best known federal public agency with fire management responsibilities in the United States (and possibly the world) is the U.S. Department of Agriculture (USDA) Forest Service. Also known less formally as the U.S. Forest Service (USFS), it administers 191 million acres (77.3 million hectares) of national forests and national grasslands, with involvement in territories such as Puerto Rico. Management directions are provided by numerous legislative acts, such as the Forest Service's Organic Act (1897) and the National Forest Management Act (1976), as well as mandates from the executive office of the president. Fire management concerns cross a myriad of fire control, fire use, ecosystem management, long-term sustainability, and fire research issues. Fire management activities are administered through the office of Fire and Aviation Management (State and Private Forestry); fire research is housed within the Research branch. In addition, the U.S. Forest Service and its federal partners are recognized as leaders in mobilization and response to emergency incidents, whether wildfires, oil spills, *Columbia* shuttle parts recovery, or natural disasters. No other civilian entity has had as much experience in gathering incident management resources and transporting teams to respond to the scene of emergency incidents. No other federal agency has had such a long track record of land-use planning.

The chief of the U.S. Forest Service is a career civil servant who oversees the entire agency, including more than 600 ranger districts, 150 national forests and 20 grasslands, 9 regions, and 6 research stations. (source: *http://www.fs.usda.gov/budget_2004/ organization.shtml*). At one time or another, all employees may be called upon to assist with and support fire operations, depending on skills, physical fitness, and training qualifications.

USDI Agencies: National Park Service, Fish and Wildlife Service, Bureau of Land Management, and Bureau of Indian Affairs

USDI National Park Service
Fire and Aviation Management
3833 S. Development Avenue
Boise, ID 83705

The other federal agencies with fire management responsibilities are housed in the U.S. Department of Interior, including the National Park Service (NPS), Fish and Wildlife Service (FWS), Bureau of Land Management (BLM), and Bureau of Indian Affairs (BIA). The BLM has a multiple use mission, while the NPS manages fire in national parks and monuments. The BIA provides liaison with tribal nations, while the FWS manages fire control and use on federal wildlife refuges.

Interior fire agencies share common fire regimes and management practices with the USFS, and all the agencies attempt to work together seamlessly—for example, on fire incident overhead teams. Some distinctions between agencies are unavoidable because of differences between Interior and Agriculture departments. For example, Interior agencies use a different fire report form (DI 1202 versus USFS 5100–29). Fire programs in the NPS are less structured generally than the USFS, perhaps out of tradition. Both the director of the NPS and the chief of the Forest Service are political appointees, though for much of the twentieth century the chief was insulated from presidential politics. More recently, elected presidents will bring in their own new leaders for both of these influential subcabinet positions.

The National Park Service (NPS) is perhaps the second most visible federal agency with fire management responsibilities, because of its responsibility for the diverse system of more than 300 national parks, monuments, preserves, and other historic and natural treasures. According to its Organic Act (1916), the NPS mandate is to "conserve the scenery and the natural and historic objects and the wild life therein and to provide for the enjoyment of the same in such manner as will leave them unimpaired for the enjoyment of future generations" (http://www.nps.gov/fire/). According to NPS guidelines, wildland fire management activities are essential to the protection of human life, personal property, and irreplaceable natural and cultural resources, as well as to the accomplishment of the NPS mission; they include both fire control and fire use activities. In fact, the National Park Service has pioneered the use of prescribed fire and natural ignitions among federal agencies since the mid-twentieth century. In the NPS, prescribed fire is centrally budgeted and administered, unlike other federal agencies.

USDI Fish and Wildlife Service
Fire and Aviation Management
3833 S. Development Avenue
Boise, ID 83705

The U.S. Fish and Wildlife Service (http://fire.r9.fws.gov/) administers the nation's wildlife refuge system, consisting of more than 500 refuges sprinkled across the country. The FWS maintains an active prescribed fire program (perhaps the largest of any federal agency in terms of total annual area treated), since refuges provide habitat for many wildlife species that live in fire-adapted ecosystems. The FWS also maintains an active suppression organization and cooperates with other federal agencies in managing complex wildfire incidents. The bulk of FWS prescribed fire activity occurs on refuge lands in the southeastern United States, although as with other federal agencies, prescribed fire treatment areas are on the increase nationwide.

USDI Bureau of Land Management
Fire and Aviation Management
3833 S. Development Avenue
Boise, ID 83705

The Bureau of Land Management (http://www.fire.blm.gov /index.htm) is the Department of Interior's multiple use agency, responsible primarily for public rangelands used for livestock grazing and also forests. The agency manages 264 million acres (109.3 million hectares) of public land across the United States and provides fire protection on 388 million acres of public and state land. However, no single federal or state fire organization is designed to handle its entire fire workload, and the BLM is no exception. By using common practices, operating and physical fitness standards, training, and closest forces concepts, the BLM is committed to interagency cooperation with its numerous partners. The agency is also committed to exchanges of protection that are a necessary and integral part of wildland fire management in the United States. In the western United States alone, the BLM has agreements with at least 1,500 wildland fire agencies. The agency also operates smokejumper facilities at the National Interagency Fire Center in Boise, Idaho, and the Alaska Fire Center in Fairbanks.

USDI Bureau of Indian Affairs
Fire and Aviation Management
3833 S. Development Avenue
Boise, ID 83705

The Bureau of Indian Affairs (BIA) is unique among federal agencies, with trust and government-to-government relationships between the United States and more than 560 tribal nations. With roots reaching back to the Continental Congress, the BIA is almost as old as the United States itself. For most of its existence, the BIA has mirrored the public's ambivalence toward the nation's indigenous peoples by carrying out federal policies that have helped or hurt native populations. But the BIA's mission has evolved, just as federal policy has changed from subjugation and assimilation of American Indian and Alaska Native peoples, to relationships based on partnership and service.

BIA administers lands held in trust for the benefit of Native American tribes, including 55.7 million acres (22.7 million hectares) of land held in trust by the United States for American Indians, Indian tribes, and Alaska Natives. Some 78 percent of lands are tribally owned and 18 percent individually owned, with the rest held in trust by the federal government. Fire management responsibilities include effective wildland fire protection, fire use and hazardous fuels management, and timely emergency rehabilitation on Indian forest and range lands held in trust by the United States, based on fire management plans approved by the Indian landowner. Preparedness will be based on the most efficient level of meeting tribal goals and objectives for the program, using Indian resources, and a cooperative, interagency approach to meeting local, regional, and national resource needs; additionally, an effective fire prevention program focuses on human-caused fires. Implementation of tribal management of the program will be facilitated under Indian Self-Determination, as requested by tribal government. Today BIA provides federal services to approximately 1.5 million American Indians and Alaska Natives who are members of more than 560 federally recognized Indian tribes in the contiguous United States and in Alaska. The past thirty years have seen the largest number of American Indian and Alaskan Native people working for the BIA—constituting about 90 percent of its 10,746 employees—more than at any other time in the agency's history (sources: http://www.bianifc.org/; http://www.doiu.nbc.gov /orientation/bia2.cfm).

Federal Emergency Management Agency (FEMA)
500 C Street SW
Washington, DC 20472
http://www.fema.gov

FEMA is a formerly independent agency that became part of the new Department of Homeland Security in response to the 9–11 terrorist attacks. It is tasked with responding to, planning for, recovering from, and mitigating against disasters. FEMA can trace its beginnings to the Congressional Act of 1803. That act, generally considered the first piece of disaster legislation, provided assistance to a New Hampshire town following an extensive fire. Since then the agency has undergone several reorganizations and mission revisions, most recently with relation to the framing of natural hazards and emergency management to encompass homeland security issues. FEMA plays an instrumental role in managing natural disasters, including wildfires, floods, tornadoes, earthquakes, and hurricanes. The U.S. Fire Administration, within the Homeland Security Department and FEMA, maintains a database of firefighter fatalities, including wildland but also structural fire protection (source: http://www.fema.gov).

National Interagency Prescribed Fire Training Center (NIPFTC)
3250 Capital Circle SW
Tallahassee, FL 32310
850–523–8630 or 866–840–3247 (toll free)
http://fire.r9.fws.gov/pftc

The mission of NIPFTC is to maintain a national interagency center of excellence for prescribed fire, with an emphasis on actual field experience. This center aims to increase skills and knowledge and to build confidence in the application of prescribed fire. Participating trainees are able to advance careers while blending field experience in prescribed burning with classroom instruction. NIPFTC coordinates prescribed fire training efforts with the Southwest Fire Use Training Academy, based in Albuquerque, New Mexico.

Environmental Protection Agency (EPA)
Ariel Rios Building
1200 Pennsylvania Avenue, NW
Washington, DC 20460
202–272–0167
http://www.epa.gov/

Although the agency is not a formal fire management organization, the regulatory functions of EPA regarding clean air and water play a significant role in determining the timing and extent of agency prescribed fires, largely through rules promulgated for enforcement by individual states. In particular, EPA develops policies for implementing the current particulate matter standards related to national ambient air quality with respect to prescribed fire and its impacts; wildfire emissions are not regulated.

Government Accounting Office (GAO)
441 G Street NW, Room LM
Washington, DC 20548
202–512–6000
http://www.gao.gov/

With skyrocketing costs for wildfire management, the U.S. Congress has taken an interest in agency plans for mitigating risks and hazards through fuel treatments. The GAO is the audit, evaluation, and investigative arm of Congress. GAO exists to support the Congress in meeting its constitutional responsibilities and to help improve the performance and ensure the accountability of the U.S. federal government to the people. The GAO examines the use of public funds, evaluates federal programs and activities, and provides analyses, options, recommendations, and other assistance to help the Congress make effective oversight, policy, and funding decisions. In particular, two reports of interest are GAO/RCED-99–65 Catastrophic Wildfire Threats (1999) and GAO-03–805 Wildland Fire Fuels Reduction (2003).

State and Local Jurisdictions

State Forestry Agencies

In the United States, individual states have separate fire organizations that cooperate with federal, private, and local jurisdictions. For example, all the Western states have active fire management organizations, usually housed in respective state forestry or related organizations. Although perhaps less visible than federal agencies, the state organizations provide important fire services to local jurisdictions, including private landowners. Some of the more active state agencies nationwide engaged in wildland fire management activities include the following:

Alaska Division of Forestry
State Forester's Office
550 West 7th Avenue, Suite 1450
Anchorage, AK 99501
907–269–8474

Arizona State Land Department
2901 W. Pinnacle Peak Road
Phoenix, AZ 85027–1002
602–255–4059

California Department of Forestry and Fire Protection
P.O. Box 944246
1416 9th St., Rm. 1505
Sacramento, CA 94244–2460
916–653–7772

Colorado State Forest Service
Colorado State University
203 Forestry Building
Fort Collins, CO 80523–5060
970–491–6303

Florida Division of Forestry
3125 Conner Boulevard
Tallahassee, FL 32399–1650
850–488–4274

Kansas Forest Service
2610 Claflin Road
Manhattan, KS 66502–2798
785–552–3300

Maine Forest Service
22 State House Station
Harlow Building
Augusta, ME 04333
207–287–2791

Minnesota Division of Forestry
500 Lafayette Road
St. Paul, MN 55155–4044
651–296–5954

Mississippi Forestry Commission
301 N. Lamar St., Suite 300
Jackson, MS 39201
601–359–1386

Montana Department of Natural Resources and Conservation, Forestry Division
2705 Spurgin Road
Missoula, MT 59804
406–542–4217

Nebraska Forest Service
Room 103, Plant Industry Bldg.
Lincoln, NE 68583–0815
402–472–2944

Nevada Division of Forestry
2525 S. Carson Street
Carson City, NV 89701
775–684–2512

New Mexico Forestry Division
P.O. Box 1948
Santa Fe, NM 87504–1948
505–476–3328

New York State Department of Environmental Conservation
625 Broadway
Albany, NY 12233–4250
518–402–9405

North Carolina Division of Forest Resources
1616 Mail Service Center
Raleigh, NC 27699
919–733–2162

Oklahoma Department of Agriculture—Forestry Services
P.O. Box 528804
Oklahoma City, OK 73152–3864
405–521–3864

Oregon Department of Forestry
2600 State Street
Salem, OR 97310
503–945–7211

Pennsylvania Bureau of Forestry
P.O. Box 8552
Harrisburg, PA 17105–8552
717–787–2703

South Carolina Forestry Commission
P.O. Box 21707
Columbia, SC 29221
803–896–8800

South Dakota Resource Conservation and Forestry
Foss Building
523 E. Capitol Avenue
Pierre, SD 57501
605–773–3623

Tennessee Department of Agriculture—Division of Forestry
P.O. Box 40627
Melrose Station
Nashville, TN 37204
615–837–5411

Texas Forest Service
301 Tarrow, Suite 364
College Station, TX 77840–7896
979–458–6600

Utah Department of Natural Resources
1594 W. North Temple, Suite 3520
Salt Lake City, UT 84114–5703
801–538–5530

Washington Department of Natural Resources
P.O. Box 47001
1111 Washington Street
Olympia, WA 98504–7001
360–902–1603

West Virginia Division of Forestry
1900 Kanawa Boulevard, East
P.O. Box 7921
Madison, WI 53707–7921
304–558–3446

**Wisconsin Department of Natural Resources—Division of
Natural Resources**
P.O. Box 7921
Madison, WI 53707–7921
608–264–9224

Wyoming State Forestry Division
1100 West 22nd Street
Cheyenne, WY 82002
307–777–7586

Within states, local county and city jurisdictions may also support
active fire organizations—for example, Los Angeles County and
Los Angeles City fire departments in southern California. Depending on location, these local fire agencies may have firefighting resources that in size and budget may rival or exceed their
federal partners. Also, in urban interface or rural areas, private
and volunteer fire departments often play a pivotal role in initial
attack operations. In fact, in some urban interface areas, volunteer
fire departments may arrive first at a fire, even before a public
agency with local jurisdiction. Volunteers may provide critical assistance on large incidents given their knowledge of local weather
patterns and prevalent fuel characteristics. Private companies
may become even more important as governments cut back on
services in response to declining budgets.

National Association of State Foresters Washington Office
444 N. Capitol St, NW Suite 540
Washington, DC 20001
202–624–5415
http://www.stateforesters.org/

States are represented on the National Wildfire Coordinating Group (NWCG) through affiliation with the National Association of State Foresters (http://www.stateforesters.org/). The National Association of State Foresters, the Forest Service, and the Ad Council oversee the Smokey Bear program.

According to their website, the state foresters provide management assistance and protection services for more than two-thirds of the nation's forests, including assistance to nearly 10 million nonindustrial private forest (NIPF) landowners who operate under a wide variety of goals and circumstances. These private forest lands provide more than 70 percent of the nation's wood supply, and contribute to many other public benefits such as clean air, clean water, healthy wildlife habitat, and outdoor recreation. In addition, state forestry agencies develop urban and community forestry programs as well as forest protection programs (to deal with wildfires, destructive pests, and diseases).

Western Governors Association
1515 Cleveland Place
Suite 200
Denver, CO 80202–5114
303–623–9378
http://www.westgov.org/

The Western Governors Association (WGA) represents eighteen Western states and three U.S. Flag Pacific islands on a variety of issues, from energy policy to forest health. One Western governor assumes leadership for a two-year term, with an executive director based in Denver, Colorado.

Following the catastrophic wildfire season of 2000, Western governors and the secretaries of the U.S. departments of Agriculture and Interior agreed to form state, federal, and local teams to develop solutions to recover forests devastated by that year's wildfires and to improve the overall health of forests to prevent future fires.

Key points made by the governors were the following:

- States should be fully represented in all prioritization, implementation, and decision-making activities at the national, state, and local levels.
- All funds provided by the Congress should be made available for expenditure across the landscape,

including federal and nonfederal ownerships for highest priority projects.

- After emergency response measures are underway, work should begin on the collaborative development of a ten-year state-federal plan for fuel reduction and forest restoration on landscapes at risk from catastrophic fire.

Nongovernmental Organizations

To date, wildland fire has been dominated by government entities such as those noted above, primarily because fire management has traditionally been handled by public agencies. But times are changing. Private sector interests in fire management grow continuously, especially with tighter budgets and personnel ceilings restricting government growth—but also because so many fires either start or spread onto private lands, requiring nongovernmental solutions. In addition, environmental interest groups continually press to make sure that their viewpoints are represented in land management decisions. The largest interest group may be the U.S. public, including people who are directly affected by wildfires and management activities. All citizens foot the bills for government fire and forest management programs. In return for paying our taxes, we expect that government managers will use the best available science and make wise decisions regarding our valued forest resources.

Space does not permit a listing and description of all the nongovernmental organizations involved in fire management. Some of the most prominent nongovernmental participants in fire management are noted in the following section, along with some promising newcomers. Mention of private companies does not imply endorsement of specific products.

Ambient Control Systems
1810 Gillespie Way, Suite 210
El Cajon, CA 92020–1094
877–840–5328
www.ambientalert.com

Ambient specializes in remote sensing technology solutions powered by light—that is, without batteries or AC power. Ambient's

FireALERT will detect a wildfire ignition signature and communicate the global positioning (GPS) coordinates to the control and dispatch center, providing an early warning system. In the homeowner model, in addition to detection and communication, FireALERT can also activate a suppression system to coat a structure and its perimeter with a fire retardant for protection against an oncoming wildfire. Ambient's energy management technology makes possible battery-free remote sensing, communications, and activation systems that can operate for more than twenty years.

Astaris LLC
810 East Main Street
Ontario, CA 91761
909–983–0772/800–682–3626
http://www.astaris.com

Phos-Chek® brand fire retardants and foam suppressants are used widely around the world by government agencies with wildland fire management responsibilities. Since 1963, Astaris LLC (formerly the chemical arm of Monsanto Company) has developed and supplied products, services, and equipment that increase the firefighting efficiency of water. Recently Astaris has expanded its product line to service municipal, rural, industrial, and military firefighters.

Bombardier Aerospace
Amphibious Aircraft
P.O. Box 6087, Station Center-ville
Montreal, Quebec, Canada H3C 3G9
514–855–5000
http://www.canadair415.com

U.S. firefighting operations have been aided recently by international assistance, including crews, overhead teams, and aircraft. Part of this growing trend toward reliance on international resources includes the use of Canadian amphibious aircraft that can deliver foam and water payloads from lakes and seas to knock down forest fire flames. Bombardier Aerospace's specialty aircraft (Canadair CL 215 and the larger Canadair CL 415) have made inroads into the U.S. and European markets, especially where fires burn forested areas close to large bodies of water. Minnesota and North Carolina state agencies have acquired these aircraft, while two Canadair 415 amphibians, owned by the

Province of Quebec, operate seasonally under a long-term lease with Los Angeles County.

Erickson Air-Crane, Inc.
3100 Willow Springs Rd.
P.O. Box 3247
Central Point, OR 97502
541–664–7615
http://www.ericksonaircrane.com

Erickson Air-Crane specializes in heavy-lifting helicopters for use in firefighting, logging, and precision placement of large construction parts. Its S-64 Helitanker can carry a payload of 20,000 to 25,000 pounds (9,000 to 11,300 kg) and can scoop water from lakes and make aerial drops of water or chemical-based retardants. In addition, the corporation was a principal sponsor for the 3rd International Wildland Fire Conference and Exhibition, held in October 2003 in Sydney, Australia.

Forest Service Employees for Environmental Ethics (FSEEE)
P.O. Box 11615
Eugene, OR 97440
541–484–2692
http://www.fseee.org/

FSEEE is a national organization aiming to hold the USDA Forest Service accountable for responsible land stewardship, including but not restricted to fire management practices on public lands. Borrowing on ideas first conceptualized by Aldo Leopold, a pioneer in land conservation, FSEEE aims to forge a socially responsible value system for the Forest Service based on a land ethic that ensures ecologically and economically sustainable resource management. FSEEE is made up of thousands of concerned citizens; present, former, and retired Forest Service employees; other government resource managers; and activists working to influence the Forest Service's basic land management philosophy.

International Association of Wildland Fire (IWAF)
4025 Fair Ridge Dr.
Fairfax, VA 22033
785–423–1818

The International Association of Wildland Fire (IAWF) is a nonprofit, professional association representing members of the global wildland fire community. The purpose of this trade association is to facilitate communication and provide leadership for the wildland fire community. IAWF is involved regularly in conferences and was a driving force for two publications still read widely today: *International Journal of Wildland Fire* and *Wildfire* magazine. The *International Journal of Wildland Fire* publishes original research; *Wildfire* magazine publishes articles of interest to field practitioners.

Keep Green Associations (Montana, Idaho, Oregon)
3802 Industrial Avenue
Coeur d'Alene, ID 83814
888–372–4042
http://www.keepidahogreen.org/

Keep Oregon Green
P.O. Box 12365
Salem, OR 97309–0365
503–945–7499
http://www.keeporegongreen.org/

Keep Montana Green Association
2705 Spurgin Road
Missoula, MT 59804–3199
406–542–4251
http://www.keepgreen.org/start.htm

Since the 1940s timber corporations in Idaho, Montana, and Oregon have partnered with local, state, and federal agencies and private citizens to prevent wildfires caused by human carelessness, focusing on education programs aimed at adults and children. The earliest association (Oregon Green Guard) was organized in 1940 with a membership of 5,000 youths, ages eight to sixteen years. The Keep Montana Green Association was formed in 1945 and chartered as a tax-exempt nonprofit organization in 1961. Keep Idaho Green started in 1946 to consolidate efforts of Jaycee clubs, state and federal forestry agencies, and a variety of private and public concerns. Each association has evolved uniquely, sometimes forging alliances with organizations (such as Boy and Girl Scouts, Junior Forest Rangers, and 4-H), or sponsoring different activities such as wildfire prevention poster competitions, athletic tournaments, or sign paintings (for example, the world's largest sign on Lucky Peak Dam in Idaho).

National Fire Protection Association (NFPA)
1 Batterymarch Park
Quincy, MA 02169–7471
617–770–3000
http://www.nfpa.org/catalog/home/index.asp

The NFPA is a private company founded in 1896 whose mission is to reduce the worldwide burden of fire and other hazards on the quality of life by providing and advocating scientifically based consensus codes and standards, research, training, and education. NFPA is concerned with reducing damage from both wildfires and structural fires, and it is a principal sponsor of the FIREWISE program and websites (along with USDA-Forest Service, the Department of Interior, the National Association of State Foresters, and the U.S. Fire Administration).

National Commission on Science for Sustainable Forestry (NCSSF)
1707 H St, NW, Suite 200
Washington, DC 20006
202–207–0007
http://www.ncseonline.org/NCSSF/

NCSSF is a multisponsor commission convened by the National Council for Science and the Environment (NCSE), a nonadvocacy nonprofit organization committed to improving the scientific basis for environmental decision-making. NCSSF's mission is to improve the scientific basis for the design, conduct, and evaluation of sustainable forest practices in the United States, including issues related to fires (wild and prescribed), forest health, and biodiversity. NCSSF sponsors work of the highest technical quality (such as symposia and workshops) that is relevant to the urgent needs of forest managers, practitioners, and policy-makers.

Sierra Club
National Headquarters
85 Second Street, 2nd Floor
San Francisco, CA 94105
415–977–5500
http://www.sierraclub.org/

The Sierra Club is one of the nation's oldest conservation advocacy groups, with roots stretching back to John Muir in 1892. Its mission includes (1) exploration, enjoyment, and protection of wild places on earth; (2) practice and promotion of responsible use of earth's ecosystems and resources; (3) education and enlistment of human resources to protect and restore the quality of natural and human environments; and (4) use of all lawful means to carry out club objectives. It is perhaps best known for its books and calendars, *Sierra* magazine, and its legal filings in behalf of environmental concerns, including nonrenewable energy sources, farm pollution, and timber harvests on public lands.

Society of American Foresters
5400 Grosvenor Lane
Bethesda, MD 20814–2198
301–897–8720
301–897–3690 (Fax)
http://www.safnet.org

The Society of American Foresters (SAF) is the national scientific and educational organization representing the forestry profession in the United States. Founded in 1900 by Gifford Pinchot, it is the largest professional society for foresters in the world. The mission of the Society of American Foresters is to advance the science, education, technology, and practice of forestry; to enhance the competency of its members; to establish professional excellence; and to use the knowledge, skills, and conservation ethic of the profession to ensure the continued health and use of forest ecosystems and the present and future availability of forest resources to benefit society. SAF is a nonprofit organization meeting the requirements of 501(c)(3). SAF members include natural resource professionals in public and private settings, researchers, CEOs, administrators, educators, and students (source: http://www.safnet.org).

Storm King Mountain Technologies
4725 Calle Alto
Camarillo, CA 93012
805–484–7267
http://www.stormkingmtn.com

Named after the mountain in Colorado where fourteen firefighters died in 1994, Storm King Mountain™ Technologies provides products using thermal insulating materials to protect firefighters and fire equipment in a fast-moving flaming front. Founded by Jim Roth, whose brother died in the 1994 burnover in Colorado, the company aims to make the best protective equipment for maximizing firefighter safety under circumstances in which training, tactics, and safety procedures might be insufficient to ensure survival on a wildfire.

Tall Timbers Research Station
13093 Henry Beadel Drive
Tallahassee, FL 32312–0918
850–893–4153
http://www.talltimbers.org/

A privately funded organization devoted to fire, wildlife, and conservation was the brainchild of Herbert Stoddard, a naturalist in the 1920s who studied the decline of bobwhite quail populations in southern Georgia and northern Florida. Based in Tallahassee, Florida, the station became a reality in 1958 on land donated by Henry Beadel, Tall Timbers Plantation owner. Originally dedicated primarily to research on fire and bobwhite quail in the Southeast, the research station rose to prominence with its annual Tall Timbers Fire Ecology conference series—a good starting point for budding fire ecologists to become familiar with the extant literature. The station's fire ecology emphasis has broadened considerably over the years to cover other fire regimes and geographic areas in North America, and an expanded set of activities such as annual field days and a fire ecology database. Also, the station has broadened its sponsorship of research topics to include other animal and plant species.

The Forest Trust
P.O. Box 519
Santa Fe, NM 87504
505–983–8992
http://www.theforesttrust.org/

The Forest Trust is dedicated to protecting the integrity of the forest ecosystem and improving the lives of people in rural communities. Founded in 1986, the trust attempts to represent a middle ground between public forest management agencies, such as the U.S. Forest Service, and environmental groups with preservation or "hands-off" interests. The trust challenges conventional forest management philosophies and provides resource protection strategies to environmental organizations, rural communities, and public agencies. The trust also provides land stewardship services to owners of private forest and range lands of significant conservation value.

The Nature Conservancy
4245 N. Fairfax Dr., Ste. 100
Arlington, VA 22203–1606
703–841–5300
800–628–6860
http://tnc.org

The Nature Conservancy's (nonprofit) informal goal is to save the last great natural places on earth. To that end, TNC has procured more than 92 million acres (37 million hectares) around the world, including 12 million acres (4.9 million ha) in the United States, for purposes of preserving plant and animal habitat, using a systematic, science-based approach to identifying candidate sites for protection—that is, its "Conservation by Design." As a private, not-for-profit company, TNC has flexibility to procure lands regardless of the owner's ideological beliefs. Two initiatives of note include its Fire Management Initiative and Fire Learning Network. The Fire Management Initiative (http://nature.org/initiatives/fire/) strives to mitigate problems caused by disruption of natural fire cycles globally. The North American Fire Learning Network (http://tnc-ecomanagement.org/Fire/) promotes development and testing of adaptive, multiarea fire management strategies in four demonstration landscapes. A similar network is developed for Latin America and the Caribbean.

The Wilderness Society
1615 M Street, NW
Washington, DC 20036
202–833–2300
http://www.wilderness.org

Founded in 1935, the Wilderness Society works to protect the wilderness lands of the United States and to develop a nationwide network of wildlands, through public education, scientific analysis, and advocacy. Staff members develop positions on a wide variety of wildland issues, including wild and prescribed fire on public lands (see, for example, Wilderness Society 2003).

Regional offices include Anchorage, Seattle, San Francisco, Boise, Bozeman, Denver, Atlanta, and Boston.

Forest Fire Research and Technology Centers

Fire research has played an important role in improving understanding of fire's role in wildlands and improving agency fire management capabilities. Long the domain of USDA Forest Service Experiment stations, universities, and isolated national parks with research staff, the quest for knowledge and understanding has expanded significantly following recent disastrous wildfire seasons (1994, 2000, 2002, and 2003). In particular, the Joint Fire Science Program and National Fire Plan have sponsored dramatic boosts in funding support for wildland fire research. In addition, several federal technology and development centers routinely develop products and provide services to the wildland fire management community.

Experiment Stations and Research Programs

USDA Forest Service Rocky Mountain Research Station
Fire Sciences Laboratory
5775 W. Hwy. 10
Missoula, MT 59802
406–329–4866
http://www.firelab.org

The federal government maintains an active research program in wildland fire sciences through its experiment stations (in particular, the Missoula, Pacific Northwest, and Riverside fire laboratories). The Missoula Fire Sciences Laboratory in Missoula, Montana, has active projects in fire behavior, fire effects, and fire chemistry. Facilities include combustion chambers and wind tunnels for conducting experimental fires.

USDA Forest Service Pacific Northwest Research Station
Pacific Wildland Fire Sciences Laboratory
3200 SW Jefferson Way
Corvallis, Oregon 97331

541–750–7265
http://www.fs.fed.us/pnw/fera/

In 2003, the Pacific Wildland Fire Sciences laboratory was reborn from the previous Seattle Forestry Sciences Laboratory. Head-quartered in Portland, Oregon, the new lab continues its long-standing emphasis on the impacts of fire on air quality and visibility, wildfire and ecology research, the effects of fire on air, the impacts of smoke on human health, and social research (rural and urban wildland interface).

USDA-Forest Service, Pacific Southwest Research Station
Riverside Forest Fire Laboratory
4955 Canyon Crest Drive
Riverside, California 92507
909–680–1500
http://www.rfl.psw.fs.fed.us/

The Riverside Fire Laboratory in southern California maintains active research projects dealing with air pollution and climate impacts on Western forest ecosystems, meteorology for fire severity forecasting, wildland recreation and urban cultures, pre-scribed fire and fire effects, and fire management in the wild-land/urban interface. The facility has traditionally provided a focal point for researching the role of fire, fire behavior, and fire management (including economics) in southern California chap-arral ecosystems.

USDA-Forest Service, Southern Research Station
P.O. Box 2680
Asheville, NC 28802
828–257–4832
http://www.srs.fs.usda.gov/

The world's first laboratory devoted exclusively to the study of forest fires was established in Macon, Georgia in 1958, in recognition of the South's unique fire problems. The lab was a coopera-tive venture between the USDA Forest Service and the Georgia Forestry Commission, a unique example of state and federal col-laboration. Although the original fire laboratory was later closed for budgetary reasons, the research station remains active with projects investigating prescribed fire, the urban-wildland inter-face, and the economics of forest protection.

USGS National Center
12201 Sunrise Valley Drive
Reston, VA 20192
703–648–400
http://www.usgs.gov/themes/Wildfire/fire.html

The U.S. Geological Survey (USGS) conducts fire-related research to meet the varied needs of the fire management community and to understand the role of fire in the landscape. This research includes fire management support, studies of postfire effects, and a wide range of studies on fire history and ecology. The USGS maintains science centers and three regional office headquarters (East, Central, and West) across the country.

Joint Fire Science Program
National Interagency Fire Center
3833 S. Development Ave.
Boise, ID 83705
208–387–5349
http://jfsp.nifc.gov/

The Joint Fire Science Research Program began after the 1994 fire season review revealed the need for basic information on fuels management that could be applied by field practitioners. The National Fire Plan was developed in response to perceived priorities in the areas of firefighting, rehabilitation, community assistance, fuels reduction, and accountability, all of which surfaced after the 2000 fire season. Both programs involve cooperation between the USDI, USDA, and the National Association of State Foresters to fund important fire research projects.

Although managers and researchers may work for the same or sibling government agencies, communication gaps often separate different branches working toward the same end. For example, the U.S. Forest Service employs both public land managers (in the national forest system) and scientists (in research stations), but those two branches may be separated by cultural and political differences, lack of a common language or vocabulary, and obstacles or barriers to the formation of effective partnerships. Programs such as the Joint Fire Science Program and National Fire Plan attempt to bridge those gaps, especially by soliciting manager feedback to establish research priorities; sometimes, however, they have limited effectiveness. One of the reasons for the gap may be the large disparity in funding between fire management and fire research (for example, USFS fire research was funded at approximately 1.6 percent of expenditures for fire management during fiscal year 2002).

The Joint Fire Science Program was established in 1998 as a partnership with six federal agencies to identify and encourage new research projects to fill in the gaps in knowledge about wildland fire and fuels. In the 1998 appropriation, Congress and the administration provided a more flexible funding authority to support efforts (primarily fuel treatments), with the goals of reducing the occurrence of uncharacteristically severe wildland fires and improving ecosystem health. In granting that funding authority, Congress expressed concern that "both the Forest Service and the Department of Interior lack consistent and credible information about the fuels management situation and workload, including information on fuel loads, conditions, risk, flammability potential, fire regimes, locations, effects on other resources, and priorities for treatment in the context of values to be protected." In response, the joint Fire Science Initiative identified four principal purposes:

- Fuels inventory and Mapping
- Evaluation of Fuels Treatments
- Scheduling of Fuels Treatments
- Monitoring and Evaluation

Starting with $4 million annually to fund competitive research proposals, the JFSP was expanded to $8 million annually following the year 2000 fire season. More than two hundred research projects have been funded since inception of the JFSP.

National Fire Plan
C/o USDA Forest Service
Auditor's Building
201 14th Street, SW at Independence Ave., SW
Washington, DC 20250
http://www.fireplan.gov/

The U.S. Congress adopted the National Fire Plan following the 2000 wildfire season, to be implemented by the U.S. Forest Service, Department of Interior, and their many cooperators. Emphasis areas include firefighting, rehabilitation and restoration, hazardous fuels reduction, community assistance and research, as well as improvements in accountability. Congress continues to demonstrate support, as evidenced by funding provided for the National Fire Plan in the FY2002 Interior and Related Agencies Appropriations Act for fiscal year 2002. More than $2.26 billion is allocated for National Fire Plan and base program funding. Allocations include $1,590,712,000 for the Forest Service and $678,421,000 for the Department of the Interior.

Technology and Development Centers

USDA Forest Service Missoula Technology and Development Center (MTDC)
5785 Hwy 10 W
Missoula, MT 59808
406–329–3900
http://www.fs.fed.us/eng/techdev/mtdc.htm

The Missoula Technology and Development Center (MTDC) began as the support facility for Forest Service fire management in the late 1940s, when a small group started developing techniques for parachuting men and cargo. In the early 1960s the center's role was expanded to a servicewide technical center with a nationwide program that now encompasses all Forest Service equipment needs. Today, MTDC works with federal and state agencies, universities, private firms, and research groups to meet its responsibilities to resource managers.

USDA Forest Service San Dimas Technology and Development Center (SDTDC)
444 E. Bonita Ave.
San Dimas, CA 91773
909–599–1267
http://www.fs.fed.us/eng/techdev/sdtdc.htm

The San Dimas Technology and Development Center (SDTDC) was established in 1945 (in Arcadia, California) to standardize fire equipment and to address fire control requirements in the West. The southern California location was selected because of the fire activity in the area, its evolving industrial and academic centers, and the availability of space in an existing Los Angeles County Forest Service facility. In 1965 a new facility, thirteen miles east of Arcadia, was designed and constructed to house the center, close to the San Dimas Experimental Forest, a natural laboratory for fire research on chaparral ecosystems.

USDA Forest Service Remote Sensing Application Center (RSAC)
222 West 2300 South
Salt Lake City, UT 84119
801–975–3750
http://www.fs.fed.us/eng/rsac
http://fsweb.rsac.fs.fed.us

Remote sensing gathers data about an object or event (such as a wildfire) from a distance—that is, airborne and satellite imagery. RSAC is colocated in Salt Lake City, Utah, with the USDA Forest Service Geospatial Service and Technology Center to provide national assistance to agency field units in applying the most advanced technology available toward monitoring and mapping of natural resources.

Education and Training

High school graduates have several options for obtaining training in wildfire management or fire science. The most direct option is to apply for employment with a public agency and receive on-the-job training. For some individuals, that path may fulfill career aspirations for the short term but may fall short in terms of advancement possibilities over the course of a career. Higher education may make more sense for those seeking the rewards of a professional career in fire management, including more challenging work and a higher starting salary. Alternatively, some may opt to attend institutions (including two-year colleges) that offer fire science programs which focus on wildland fire as well as structural fire protection, law enforcement, and emergency medical technician training. Examples in Colorado include Aims, Aurora, Front Range, and Red Rocks community colleges.

The diversity of university programs that include the study of forest fires mirrors the large variety of higher education institutions in the United States. Thus there are many ways for interested university students to broaden their understanding of fire science. By and large, most of the accredited university forestry programs (accredited, that is, by the Society of American Foresters) include one or more courses dealing with fire ecology or fire management—although many will immerse the study of fire within team-taught, multidisciplinary courses. Only a handful of university programs may offer a specific set of fire courses, and even fewer will offer a concentration or allow students to specialize in fire studies.

Faculty members at many universities and colleges nationwide carry out active programs in wildland fire research, which can be incorporated into courses that allow students to keep up with the latest knowledge and enhance learning opportunities. Thus, even if specific programs or courses are not offered in fire science, a student may be able to study fire through independent study with a faculty member who shares that interest. A list of four-year and two-year instructional programs is available at http://www.wildlandfire.com/docs/firesci_edu.htm. Below is a sampling of universities and colleges with forestry programs in the western United States where students may learn about wildland fire science:

California Polytechnic State University, San Luis Obispo
College of Agriculture
San Luis Obispo CA 93407
805–756–2968
http://www.calpoly.edu/

Cal Poly offers an undergraduate student major in forestry and natural resources that includes emphasis areas in watershed, chaparral, and fire management, and fire and fuels management, along with several other study options. A master's degree in a related field can be earned through the College of Agriculture.

Colorado State University
College of Natural Resources
Fort Collins, CO 80523
970–491–6675
http://www.cnr.colostate.edu

Colorado State University offers a forest fire science concentration under the undergraduate forestry major. Other undergraduate options within the forestry major include forest management, forest biology, and forestry business. Students majoring in the less structured natural resource management degree may take fire courses as electives. Graduate degrees include the master of forestry (M.F.), M.S., and Ph.D. in forest fire science.

Humboldt State University
Department of Forestry and Watershed Management
1 Harpst Street
Arcata, CA 95521–8299
707- 826–3935
http://www.humboldt.edu/~for/index.shtml

One of the most recent additions to wildland fire science instructional programs is at Humboldt State University. Undergraduates earn the forestry degree, in preparation for a professional career. Graduate students can earn the master's in Natural Resources.

Northern Arizona University
School of Forestry
P.O. Box 15018
110 E Pine Knoll Drive
Flagstaff, AZ 86011–5018
928–523–3031
http://www.for.nau.edu/

NAU offers a unique program in forestry that involves a holistic approach to the environment in the junior and senior years, rather than specialty courses. Graduate degrees include a master of science in forestry (M.S.F.) and Ph.D. in forest science.

Oklahoma State University
College of Agricultural Sciences and Natural Resources
Stillwater, OK 74078
405–744–5000
http://dasnr.okstate.edu/casnr/NewCASNR/home.htm

Four undergraduate degree options are offered: forest management, natural resources conservation and management, forest ecosystem science, and urban and community forestry. For graduate students interested in advanced fire or forestry studies, the university offers a master's degree in forest resources (or environmental sciences) and Ph.D. (environmental sciences or plant sciences).

Oregon State University
College of Forestry
Corvallis, OR 97331
541–737–2004
http://www.cof.orst.edu/

The College of Forestry at Oregon State offers several undergraduate degrees including forest management and natural resources. Graduate programs leading to the master of forestry (M.F.), master of science (M.S.), or Ph.D. degrees are offered in four departments of the College of Forestry.

University of California (Berkeley)
College of Natural Resources
Department of Environmental Sciences, Policy, and Management
145 Mulford Hall # 3114
Berkeley, CA 94720–3114
510–642–6410
http://espm.berkeley.edu/

The Department of Environmental Sciences, Policy and Management offers undergraduate degrees in forestry, conservation and resource studies, and resource management. Ph.D. and master's degrees in environmental science, policy, and management; the M.F. in forestry; and the M.S. in range management are available.

University of Idaho
College of Natural Resources
Moscow, ID 83844
208–835–2397
http://www.cnr.uidaho.edu/forres/

The Department of Forest Resources offers programs leading to the following degrees: bachelor of science in forest resources, with options in administration, forest ecosystem management, and science; bachelor of science in natural resources ecology and conservation; master of science (thesis and nonthesis options); and Ph.D. with a major in natural resources (administered at the college level for all departments).

The University of Montana
College of Forestry and Conservation
Missoula, MT 59812
406–243–5521
http://www.forestry.umt.edu/academics/default.htm

The College of Forestry and Conservation offers undergraduate degrees in forest resources and resource conservation, among others. Graduate degrees include the master of science and Ph.D. in forest science.

University of Washington
College of Forest Resources
Box 352100
Seattle, WA 98195–2100
206–543–2730
http://www.forestry.umt.edu/academics/default.htm

The College of Forest Resources offers undergraduate degree programs in forest management and conservation of wildland resources. Graduate degree programs include the master of forest resources, master of science, and Ph.D.

Utah State University
College of Natural Resources
5200 Old Main Hill, Logan UT 84322
435–797–2445
http://www.cnr.usu.edu/frws

The Department of Forest, Range, and Wildlife Resources offers undergraduate degrees in forestry, and M.S. and Ph.D. degrees.

Community College Programs

The above list focuses on four-year university resident instruction programs, although universities are increasingly making use of web-based instruction and may offer distance-learning options as well. Other options for obtaining applicable knowledge and skills include community colleges, agency training institutes, and private organizations.

Many community colleges across the United States offer associates degrees or certification programs in fire science, including wildland and structural fire protection programs, also linked to fields such as hazardous materials management, law enforcement or emergency medical technician or paramedical training. The list of available options is constantly changing, but it may be accessed at http://www.wildlandfire.com/docs/firesci_edu .htm. Specific details are available at each community college, but a sampling of just a few offerings is provided below. Your local community college can be contacted to provide information on specific opportunities.

Aims Community College
Fire Science Technology
Trades and Industry Building, Room 105
5401 W. 20th Street
P.O. Box 69
Greeley, CO 80632
970–330–8008, Ext. 6485
http://www.aims.edu/academics/fire_science/index.htm

Casper College
125 College Drive
Casper, WY 82601
307–268–2110
1–800–442–2963
http://www.caspercollege.edu/trades/fire/

Lassen Community College
Fire Science
P.O. Box 3000
Susanville, CA 96130
530–257–6181
http://www.lassencollege.edu/coursesfirescience.html

Red Rocks Community College
Fire Science Technology
13300 West Sixth Ave
Lakewood, CO 80228–1255
303–914–6600
http://www.rrcc.cccoes.edu/fire/FireAcademyp.htm

Sierra College
Fire Technology Vocational Education
5000 Rocklin Rd
Rocklin, CA 95677
916–624–3333
http://www.sierracollege.edu/ed_programs/voc_edu/firetech.
html

Utah Valley State College
Utah Fire and Rescue Academy
800 West University Parkway
Orem, UT 84058
801–863-INFO
http://www.uvsc.edu/ufra/training.html

Training/Education Opportunities

Government employees generally seek advanced training to stay abreast with the latest developments in fire management as well as to obtain certification to perform certain tasks within the fire management organization. Some agencies track fire qualifications to make sure that employees stay current and are able to perform tasks with required levels of competency. In other agencies, employees may keep a "task book" that identifies training needs requisite to performing future tasks on the job. Some of these training needs can be met by courses offered at colleges and universities, although most academic institutions are not in a position to offer the knowledge and skills necessary to manage, for example, a large fire organization.

Fire-related knowledge and skills can be obtained from agency "wildfire training institutes," or sometimes from formal apprenticeship opportunities offered in some states. For example, Colorado offers an annual wildfire training institute in which participants can sign up for a variety of courses, from the most elementary "S190: Basic Firefighter" course to fire engine pump operation or incident command system. Larimer County in Colorado offers the "basic firefighter" course as part of its annual recruitment of personnel for its stand-by firefighting crews. California is developing a Fire Studies Institute, to be offered at a former military base that has been renovated to provide advanced fire study opportunities.

Washington Institute, Inc.
P.O. Box 1108
Duvall, WA 98109
425–788–5161
http://www.washingtoninstitute.net/

Rather than providing training for specific job assignments, the mission of the Washington Institute (WI) is to provide an educational enhancement experience to midcareer fire and fuels managers with public agencies. In cooperation with Colorado State University, WI offers the highly regarded Technical Fire Management (TFM) series of instructional modules for practitioners. TFM has been offered for three decades, beginning in 1981, providing managers and technicians from federal and state agencies nationwide with knowledge and tools to apply the most current fire-related technology to the management of different fire regimes. The modules include (1) Math fundamentals; (2) Statistics; (3) Economics; (4) Fuels Management; (5) Fire Effects; (6) Fire and Land Management; and (7) TFM project. The final project allows participants to apply concepts from all of the previous instructional modules to the solution of a management problem on their home unit. Periodically WI also offers stand-alone educational offerings in fuels management and fire ecology.

National Advanced Fire and Resource Institute
3265 East Universal Way
Tucson, AZ 85706
520–779–8787
http://www.nafri.gov/

Formal courses are offered for higher-level fire management decision-makers who work on extremely complex incidents. Although these training opportunities may not be open to the general public, a quick scan of the courses offered may be instructive because the listing indicates some of the skills and aptitudes required to do various jobs within the fire management organization. For example, the National Advanced Fire and Resource Institute (NAFRI) in Tucson, Arizona, offers courses in the following subject areas (source: http://nartc.net/2002/index.htm):

> *ARAU Aerial Retardant Application and Use:* Designed primarily for incident personnel on the ground, who are making the aerial retardant application decisions and managing the results. Safe, efficient, and cost-effective use of aerial retardants is stressed.
>
> *L-480/L580 Interagency Management Team Leadership:* (In development)
>
> *D-510 Supervisory Dispatcher:* Designed to train individuals in the function and responsibilities of a supervisory dispatcher within an incident support organization. The course will provide trainees with a working knowledge of the necessary management skills and operational procedures to perform the job successfully.

M-580 Fire in Ecosystem Management: This course (1) provides a conceptual framework for understanding ecosystem management; (2) explores the role of fire and fire management; (3) examines social, political, legal, economic, and environmental factors; (4) presents real-world examples of fire management applications; (5) presents the perspective of land managers on ecosystem management and the participant roles; and (6) provides participants with the tools, ideas, concepts, techniques, and methodologies for addressing fire and ecosystem management issues at their home unit.

FML Fire Management Leadership: A comprehensive look at the agency administrator's leadership role within the fire management program, including the new Federal Wildland Fire Management Policy. The complex fire situation is highlighted with the intent to provide a comprehensive understanding of processes necessary to effectively manage the situation. Included are the agency administrator's briefing, wildland fire situation analysis, risk and cost analysis, and other critical items. Policy, authority, and responsibility are clarified, discussed, and reinforced using case examples.

NAFA National Aerial Firefighting Academy: The National Aerial Fire Fighting Academy (NAFA) is a training effort by and for personnel directly involved in aerial retardant delivery.

Advanced National Fire Danger Rating System: This advanced course focuses on the role of fire danger rating in resource decision-making, through development of a fire danger rating operating plan. It provides skills for analyzing and troubleshooting fire danger rating from multiple scales. It also provides individuals with the technical skills to serve as instructors for regional courses for all wildland fire agencies using NFDRS.

NFMAS National Fire Management Analysis System: Links resource management and budget actions and fire management analytical techniques, describing how the analytical process works and how to use its outputs. An overview of the National Fire Management Analysis is covered, along with a review of programs available, the data needed, the analysis process, links with fire planning, unit resource planning, and budgeting.

Rx540 Applied Fire Effects: A science-based course designed to support the integration of fire effects knowledge into a wildland fire use program. Emphasizes the importance of fire effects in the design, implementation, and monitoring of fire treatments over multiple spatial and temporal scales. Recognizes that planning and implementing fire use are interdependent activities, and provides opportunities for burn boss and planner to work together.

S-520 Advanced Incident Management: Using classroom lecture and simulated incidents, the student will understand the role and function of a Type I Incident Management Team, the applicability of management principles to the incident management job, and the special considerations of incident management within geographic areas of the nation.

S-580 Advanced Fire Use Applications: Provides participants with an understanding of the implementation process for appropriate management response to accomplishing resource benefits over a wide range of management situations.

S-590 Advanced Fire Behavior Interpretation: Through the use of lecture, demonstrations, and exercises, the students will learn the job of the fire behavior analyst. Units covered include Fire Behavior Documentation Package, Legal Considerations, Safety, Briefings, Forecasts, Decision Support, Plan Development, and Fire Operations. Knowledge gained includes organizational, communication, and computer skills.

S-620 Area Command: Using classroom lecture and simulated incidents, the student will understand principles and concepts of Area Command and other incident organizations, agencies, and political entities.

SLAM Senior Level Aviation Management: Provides senior managers that direct or support natural resource programs with an overview of successful aviation program characteristics, with an emphasis on laws, regulations, and policies governing public and civil aviation operations. Concepts, philosophies, and theories of safety management and the leadership role in policy implementation are also addressed.

M-581 Fire Program Management: This course is designed to provide participants with the ability to apply fire management principles in assuming the duties and responsibilities of fire program management using sound decision-making, personal accountability, identifying reference resources, and applying state-of-the-art tools and methods.

Global Agencies

Forest and grassland fire effects extend to continental and intercontinental scales, as for example when smoke plumes from large wildfires contribute to reduced visibility and haze throughout a global region, as experienced in southeast Asia during 1997–1998 and the Russian Federation in 1998 and 2002. In addition, nations may call on one another for assistance when large, complex fires exceed the management capabilities of a single government. In fact, as of October 2003, twenty-one different forest fire alerts had been issued during the year by the Food and Agriculture Organization (FAO) of the United Nations, indicating threats to people and property with serious land degradation and food security. The twenty-one national fire alerts covered fire situations on every continent except Antarctica, including seven alerts in Europe, five in North America, four in Asia, three in South America, and one each in Africa and Australia.

The level of international cooperation in managing wildland fires is in its infancy, though it is growing through the efforts of organizations such as the UN. In fact the first inventory of fire activity worldwide, including savanna and agriculture burning, was only recently completed (FAO 2001). Major conclusions from the assessment included the following:

- Drought years in the 1990s caused widespread burning in tropical rain forests with significant impacts on natural resources, public health, transportation, navigation, and air quality.

- Many countries and regions have well-developed systems for documenting, reporting, and evaluating wildfire statistics in a systematic manner, but they may lack information on the effects of these fires. Many other countries do not yet have such a system, largely because of more pressing social issues. Satellite systems have been used effectively to map active fires and burned areas, especially in remote areas where other damage assessment capabilities are not available.
- Even those countries supporting highly financed fire management organizations are not exempt from the ravages of wildfires in drought years. When wildland fuels have accumulated to high levels, no amount of firefighting resources can make much of a difference until the weather moderates (as observed in the United States in the 2000 fire season).
- Uncontrolled use of fire for forest conversion, agricultural, and pastoral purposes continues to cause a serious loss of forest resources, especially in tropical areas. Some countries are beginning to realize that intersectoral coordination of land use policies and practices is an essential element in reducing wildfire losses. There were numerous examples in the 1990s of unprecedented levels of intersectoral and international cooperation in helping to lessen the impact of wildfires on people, property, and natural resources.
- Examples exist in which sustainable land use practices and the participation of local communities in integrated forest fire management systems are being employed to reduce resource losses from wildfires.
- In some countries, volunteer rural fire brigades are successful in responding quickly and efficiently to wildfires within their home range, and residents are taking more responsibility to ensure that homes will survive wildfires.
- Although prescribed burning is being used in many countries to reduce wildfire hazards and achieve resource benefits, other countries have prohibitions against the use of prescribed fire.
- Fire ecology principles and fire regime classification systems are being used effectively as an integral part of resource management and fire management planning.

- Fire research scientists have been conducting cooperative research projects on a global scale to improve understanding of fire behavior, fire effects, fire emissions, climate change, and public health.
- Institutions like the Global Fire Monitoring Center have been instrumental in bringing the world's fire situation to the attention of a global audience via the Internet (source: International Wildland Fire Summit 2003).

A global framework has been adopted for developing regional networks to link managers, policy-makers, technical experts, and scientists for purposes of early warning, fire monitoring and impact assessment, and other cooperative efforts (Mutch 2003). The extent to which regional networks have been implemented varies, but each has unique fire problems and infrastructure for attempting solutions. Regional Wildland Fire Networks (facilitated through the United Nations), after Mutch (ibid.), include the following:

- South East Asia (ASEAN)
- Australasia
- Baltic
- Central Asia
- Mediterranean
- Balkan
- Sub-Saharan Africa
- North America
- Meso-America
- South America

In principle, the vision of global or regional cooperation makes good sense, inasmuch as fire effects can transcend national boundaries, as for example between the United States and Canada. This recognition gave rise to the first International Wildland Fire Summit held in Sydney in 2003 and attended by representatives from thirty-four countries and ten international organizations. In reality, global cooperation within and among networks and nations will require many years to implement because of cultural and structural differences between nations, even within the same global region. Even so, examples of international cooperation in fire management have been in existence for some time and appear to be on the increase. For example, the Fire Management Working Group of the North American Forestry Commission was established in 1962 and has allowed exchange of firefighting resources, knowledge, and technology among Canada, the United States, and Mexico. Similar exchanges have been arranged between Australia and the United States and throughout Australasia (ibid.).

Food and Agriculture Organization of the United Nations (FAO)
Viale delle Terme di Caracalla, 00100 Rome, Italy
+39 06 5705 1
http://www.fao.org/

FAO supports international cooperation related to large wildfire incidents that threaten life and property. Alerts are issued to inform the international community about significant fire events occurring worldwide with summary reports and links to satellite imagery via the Global Fire Monitoring Center (http://www.fire.uni-freiburg.de/).

The International Tropical Timber Organization (ITTO)
International Organizations Center, 5th Floor, Pacifico-Yokohama
1–1–1 Minato-Mirai, Nishi-ku
Yokohama, 220–0012 Japan
81–45–223–1110
http://www.itto.or.jp

ITTO brings together fifty-seven member states, including thirty-one from the tropics, to cooperate on forestry and fire problems. For example, the 1997–1998 fires in Borneo and Sumatra that burned 5 million hectares mobilized ITTO missions to Indonesia and Malaysia. Other efforts have focused on fire problems in Ghana and Côte d'Ivoire.

ATSR World Fire Atlas
European Space Agency—ESA/ESRIN
via Galileo Galilei, CP 64,
00044 Frascati, Italy.
http://shark1.esrin.esa.it/ionia/FIRE/AF/ATSR/

Maps showing monthly areas burned globally since 1995 are being compiled experimentally, of special interest to researchers interested in the possible impacts of biomass burning on atmospheric processes, chemistry, or other coarse-scale assessments.

Agencies and Organizations around the World

This chapter concludes with a few examples of agencies and organizations with fire-related interests in various parts of the globe, with focus on areas with fire problems similar to those of the United States. In reality, each country on the globe confronts a unique set of fires, fire problems, and attempted solutions. The agencies indicated below represent just a sampling of the myriad organizations attempting to cope with fire globally.

Australia

Bushfire Cooperative Research Centre (CRC)
Chief Executive Officer, Bushfire CRC
5th Floor, 340 Albert St.
East Melbourne, Victoria, AUSTRALIA 3002
61–3–9412 9600
http://www.bushfirecrc.com

The Bushfire CRC was formed in 2003 following the disastrous fires in eastern Australia in 2001–2002. The organization aims to better manage bushfire risk in Australia, relying on a multidisciplinary collaboration among expert fire authorities from Australia and New Zealand, the Commonwealth Scientific and Industrial Research Organization (CSIRO) of Australia, universities, the Commonwealth Bureau of Meteorology, and others—some thirty partners in all.

Commonwealth Scientific and Industrial Research Organization (CSIRO)
CSIRO Forestry and Forest Products
P.O. Box E4008
Kingston ACT 2604
02 6281 8341
http://www.ffp.csiro.au/fap/fire.html

The Forestry and Forest Products Division at CSIRO carries out fire behavior and management research in Australia. CSIRO is a large, multidisciplinary, and diverse research organization devoted to developing innovative solutions to problems of industry, society, and the environment in Australia and on the globe. The efforts of the fire behavior and management team are aimed at reducing loss of life and property during bushfires through a better understanding of how bushfires spread and react to changes in the environment, with special focus on fire behavior in dry-eucalypt forests. CSIRO publishes *The International Journal of Wildland Fire*.

Fire Protection Association Australia (FPA Australia)
FPA Australia National Office
13 Ellingworth Parade, Box Hill, Victoria (Australia)
P.O. Box 1049, Box Hill, VIC, 3128, Australia
+61 (0)3 9890–1544
http://www.fpaa.com.au/

Fire Protection Association Australia is the country's largest organization (with 1,600 members) seeking to promote fire awareness and efforts of the fire protection industry. Aims are met through publications, conferences, exhibitions, seminars, workshops, sales of promotional materials, and coordination of Fire Awareness Week between states, among other activities.

New South Wales Rural Fire Service (RFS)
Unit 3, 175–179 James Ruse Drive, Rosehill
Locked Mail Bag 17
Granville, NSW 2142 Australia
02–9684–4411
http://www.rfs.nsw.gov.au

The New South Wales Rural Fire Service comprises about 2,200 rural fire brigades with almost 70,000 volunteer firefighters; it is responsible for fire management over approximately 95 percent of the area of New South Wales, Australia. The RFS is the world's largest fire service, protecting some of the most fire-prone areas on the planet while providing services in fire control, operations, strategic development, and risk management.

Nature Conservation Council of New South Wales
Bush Fire Management Program
Level 5, 362 Kent Street, Sydney NSW 2000
02 9279 2120 Fax: 02 9279 2499
http://www.nccnsw.org.au/bushfire/

The Bushfire Program carries out management programs aimed at sustaining natural resources and protection of life and property, within the auspices of the Nature Conservation Council (NCC) of New South Wales. Since its inception, the NCC has held conferences, seminars, and workshops and produced a number of publications such as newsletters, proceedings, and guidelines to communicate with the public.

Canada

Canadian Forest Service
Pacific Forestry Centre
506 West Burnside Road
Victoria, British Columbia
V8Z 1M5 CANADA
250–363–0600
http://www.pfc.cfs.nrcan.gc.ca/fires/

Northern Forestry Centre
5320—122nd Street
Edmonton, Alberta
T6H 3S5 CANADA
780–435–7210
780–435–7359 (Fax)
http://nofc.cfs.nrcan.gc.ca/index_e.html

Great Lakes Forestry Centre
1219 Queen Street East
Sault Ste. Marie, Ontario
P6A 2E5 CANADA
705–949–9461
http://www.glfc.cfs.nrcan.gc.ca/index_e.html

The Canadian Forest Service promotes the sustainable development of Canada's forests and the competitiveness of the Canadian forest sector. Fire research is carried out at several centers within CFS, including the Pacific Forestry Centre (British Columbia), Northern Forestry Center (Alberta), and Great Lakes Forestry Center in Ontario (source: http://www.nrcan.gc.ca/cfs-scf /science/resrch/forestfire_e.html). The Northern Forestry Center (NOFC) has spearheaded the International Crown Fire

Experiment, one of the most important examples of interdisciplinary forest fire research in recent times. NOFC also oversees the Wildland Fire Information Centre (http://cwfis.cfs.nrcan.gc.ca /en/index_e.php), which provides users with access to **The Canadian Wildland Fire Information System (CWFIS)**, a computer-based system that monitors fire danger conditions across Canada.

Parks Canada National Office
25 Eddy Street
Hull, Quebec
Canada
K1A 0M5
888–773–8888
http://www.pc.gc.ca/progs/np-pn/ecosystem/ecosystem5_E.asp

Just as in other countries of the world, fires play an important role in the national parks of Canada. National parks are located on the Atlantic, Pacific, and Arctic coasts, and across the interior mountains, plains, and Great Lakes, reaching as far north and south as Canada extends. They range in size from just under 9 square kilometers (St. Lawrence Islands National Park of Canada) to almost 45,000 square kilometers (Wood Buffalo National Park of Canada). And they include world-renowned names such as Banff and Jasper, as well as the more recently established Ivvavik and Vuntut.

Canadian Interagency Forest Fire Centre
210–301 Weston Street
Winnipeg MB Canada
R3E 3H4
204–784–2030
http://www.ciffc.ca/

The Canadian Interagency Forest Fire Centre (CIFFC) provides operational fire-control services, as well as management and information services to its member agencies. In addition to coordinating services for all of the provinces and territories, CIFFC often coordinates the sharing of resources with the United States and other countries.

New Zealand
Forest Research
Sala Street
Private Bag 3020
Rotorua
New Zealand
+64 7 343 5899
http://www.forestresearch.co.nz/fire
(Other offices in Christchurch and in Australia)

Every year in New Zealand, approximately 2,000 wildfires burn through some 7,000 hectares of rural lands. Predicting where wildfires are most likely to break out, what fuels them and helps them burn, and how rural fire managers can be best prepared to fight them is the aim of the Forest and Rural Fire Research program. The program has two broad objectives: (1) To reduce the number and consequences of wildfires; and (2) To promote the effective use of fire as a management tool.

Europe

ISDR Interagency Task Force Working Group on Wildland Fire
Global Fire Monitoring Center (GFMC)
Fire Ecology Research Group
Max Planck Institute for Chemistry
C/o Freiburg University
79085 Freiburg, Germany
49–761–808011
http://www.fire.uni-freiburg.de/

Although Europe's fire problems are generally concentrated in the Mediterranean region (that is, Spain, Portugal, France, Italy, and Greece), much of the interest in transnational fire concerns has been stimulated through efforts of the Global Fire Monitoring Center, based in Freiburg, Germany. The GFMC has been a focal point and clearinghouse for global fire information as part of the UN's International Strategy for Disaster Reduction (ISDR). The GFMC maintains active links to fire programs within the European Commission as well as to the rest of the global fire community.

Literature Cited

FAO. 2001. "FRA Global Forest Fire Assessment 1990–2000." Forest Resource Assessment Programme, Working Paper 55, pp. 189–191. Rome: FAO, 495 p.

International Wildland Fire Summit. 2003. "An Overview of Vegetation Fires Globally." Background paper (http://www.fire.uni-freiburg.de/summit–2003/Summit%20Background%20Paper%20Global%20Situation.pdf).

Mutch, R. W. 2003. "Key for Successful International Cooperation: A Dream, a Team, and a Theme." In *Proceedings of the 3rd International Wildland Fire Conference and Exhibition*, October 3–6, 2003, Sydney, Australia. CD published by 3rd International Wildland Conference and Exhibition.

Wilderness Society. 2003. "Restoring Balance to Wildland Fire Policy." A Backgrounder Pamphlet from The Wilderness Society. Washington, DC: The Wilderness Society. 10 p.

7

Print and Nonprint Resources

Interest in wildland fire science and management has expanded tremendously in recent years, perhaps reflecting high-profile wildfires, increased coverage in the news media, and growth of the worldwide web, among other causes. Numerous print and nonprint resources are now available, covering the gamut from scientific books and journal articles, to informal "war" story articles that glorify firefighting, to fire photos and interactive educational tools and games. The drama of firefighters and their personal lives provide good subject matter for novels and movies as well. In this chapter we examine the major information resources available for interested students, lay publics, and fire groupies. I have tried to capture the highlights that might be most useful for writing reports or getting a quick overview from resources available on the web.

Print Resources

Print resources include books, peer-reviewed scientific journals, theses/dissertations, government documents (including conference proceedings), and popular magazines and mass media sources. Newspapers often cover major fire events with technical articles on fire equipment, firefighting, or human-interest pieces. Scientific journal articles report results from original research, where new knowledge is generated or synthesized about subjects as diverse as fire ecology, fire effects on organisms, fire behavior, fuels management, air tanker payload deliveries, or fire economics, to name just a few. Scientific studies generally include synthesis of existing or antecedent studies, objectives, description of methods used, results, discussion, and conclusions, followed by literature citations. Data and information from peer-reviewed journals are most reliable in terms of scientific accuracy, having been conducted with the scientific method and subjected to anonymous review by subject matter experts. Theses and dissertations developed by graduate students supervised by faculty advisers at colleges and universities also provide results from original research. However, much of the information of interest to lay publics may not be addressed in the scientific literature, which to the uninitiated may appear laden with technical detail and unfamiliar nomenclature. Conversely, sometimes the popular literature must be filtered for bias and unsubstantiated claims.

Books

For an excellent introduction to the study of fire, Steve Pyne's classic *Fire in America* is a logical starting point. He provides a historical primer on the evolution of thought and action regarding wildland fire, while chronicling important events, personalities, and institutions that have influenced fire attitudes, including pre-settlement use of fire by Native Americans. This book established Steve as a premier writer on U.S. fire policy and management.

Pyne, S. J. 1982. ***Fire in America: A Cultural History of Wildland and Rural Fire.*** Princeton: Princeton University Press, 654 p.; paperback edition, University of Washington Press, 1997.

For a narrower focus on fire, particularly on wildland firefighting, Norman Maclean's "Young Men and Fire" provides insight into life and death struggles of the smokejumpers who battled the 1949 Mann Gulch fire in Montana. Maclean explores the motivations of people who choose to jump out of perfectly good airplanes as a vocation or avocation, but also provides perspective on the U.S. Forest Service as a public agency that manages fire as part of its overall mission. This book (also available on audio tape) was completed posthumously by Maclean's son, John. John Maclean later explored the circumstances surrounding the deaths of fourteen firefighters on the South Canyon fire near Glenwood

Springs, Colorado, during 1994. Both books illustrate how individual wildfires present interesting case studies, without the technical details of more formal investigations.

Maclean, N. 1992. *Young Men and Fire.* Chicago: University of Chicago Press, 301 p.

Maclean, J. N. 1999. *Fire on the Mountain: The True Story of the South Canyon Fire.* New York: William Morrow, 275 p.

Numerous textbooks have been written on the subjects of fire management and fire ecology. Most are quite technical and aimed primarily at university students or practicing professionals. Fuller's paperback is the least technical and most easily understood by lay readers.

Fuller, M. 1991. *Forest fires.* New York: John Wiley and Sons, 238 p.

Agee, J. K. 1993. *Fire Ecology of Pacific Northwest Forests.* Washington, DC: Island Press, 493 p.

Pyne, S. J., P. L. Andrews, and R. D. Laven. 1996. *Introduction to Wildland Fire: Fire Management in the United States.* New York: John Wiley and Sons, 455 p.

Whelan, R. J. 1995. *The Ecology of Fire.* Cambridge: Cambridge University Press, 346 p.

Scientific journals (nearly all of which are accessible on-line, at least for most recent volumes) keep pace with the latest developments in fire science and management. Journals noted below include one serial devoted completely to wildland fire and numerous others in which fire papers may be found. Peer reviewed journal articles are generally most objective and less apt to include biased viewpoints.

Peer Reviewed Journals

(General focus is on forest science, forest ecology, or management, with occasional fire papers, unless otherwise noted.)

Canadian Journal of Forest Research, published in English and French by the National Research Council Canada (since 1971).

Conservation Biology, published by Blackwell Scientific Publications (since 1987).

Ecology, the official publication of the Ecological Society of America (since 1920). The society also publishes *Ecological Applications* (since 1991).

Environmental Management, published by Springer-Verlag (since 1976).

Forest Ecology and Management, published by Elsevier Scientific Publishing Company (since 1976/1977).

Forest Ecology and Management, published by Springer-Verlag (since 1976).

Forest Science, published by the Society of American Foresters (SAF) (since 1955). SAF also publishes the *Journal of Forestry* (since 1902), and regional journals: *Western Journal of Applied Forestry* (since 1986), *Southern Journal of Applied Forestry* (since 1977), and *Northern Journal of Applied Forestry* (since 1984).

International Journal of Wildland Fire, published by CSIRO Publishing (since 1991, exclusive focus on wildland fire).

Journal of Environmental Management, published by Academic Press (since 1973).

National Geographic, official journal of the National Geographic Society (since 1888).

Science, published by the American Association for the Advancement of Science (since 1883).

Technical Reports and Research Papers

In addition to the above serial publications, the USDA Forest Service publishes numerous scientific manuscripts related to fire through its experiment stations, primarily disseminated as general technical reports or research papers. The Rocky Mountain Research Station, Pacific Southwest Research Station, and Pacific Northwest Research Station most often publish fire-related papers, or often will list documents published by station scientists that are available elsewhere. Additional fire-related sources may be found (though less frequently) through the Northcentral and Southern Experiment Stations. An interested reader may contact individual research stations and request to be included on quarterly mailing lists for alerts on latest publications. The following list describes several exemplary papers from the extant literature.

Rothermel, R. C. 1983. *How to Predict the Behavior of Forest and Rangeland Fires.* USDA Forest Service General Technical Report INT-143. Ogden, UT: U.S. Department of Agriculture, Forest Service, Intermountain Research Station. 161 p. Most fire behavior prediction systems in the United States rely on a mathematical model developed in 1972 by Richard C. Rothermel. The model predicts with reasonable precision the spread rate of a fire that has reached a sustained growth rate in uniform fuels in flat or mountainous terrain. In 1983, Rothermel summarized techniques for estimating the behavior of free-spreading wildland fires under one cover.

Butler, Bret W., Roberta A. Bartlette, Larry S. Bradshaw, Jack D. Cohen, Patricia L. Andrews, Ted Putnam, and Richard J. Mangan, 1998. *Fire Behavior associated with the 1994 South Canyon Fire on Storm King Mountain, Colorado.* Research Paper RMRS-RP-9. Ogden, UT: U.S. Department of Agriculture, Forest Service, Rocky Mountain Research Station, 82 p. HTML Version: http://www.fs.fed.us/rm/pubs/SouthCanyon/index.html. PDF Version: http://www.fs.fed.us/rm/pubs/rmrs_rp09.pdf.

Actual fire behavior in the field sometimes bears little resemblance to computerized fire behavior predictions. After the fourteen firefighter fatalities during the South Canyon fire in July 1994, fire scientists assessed circumstances and suggested guidelines that could help firefighters avert such a tragedy in the future. Essentially, firefighters were caught off guard by a fire that transitioned from a relatively slow-spreading, low-intensity surface fire to a high-intensity, fast-spreading fire burning through the entire fuel complex, surface to crown. The analysis includes a detailed chronology of fire and firefighter movements, changes in the environmental factors affecting the fire behavior, and crew travel rates and fire spread rates.

Fire Effects: The "Rainbow" Series

In the 1970s the USDA Forest Service research stations sponsored a massive effort to summarize the status of knowledge on fire effects on flora, fauna, air, soils, water, and fuels. Dubbed the "rainbow series" because of the different cover colors chosen for each volume, the collection represented an important summary of fire effects knowledge at that time. Following its establishment in 1998, the Joint Fire Science Program commissioned an effort to update knowledge of fire effects. Three updates have been published so far, with the rest scheduled for future release.

Fire Effects on Fauna

Smith, J. K., ed. 2000. *Wildland Fire in Ecosystems: Effects of Fire on Fauna.* USDA Forest Service General Technical Report RMRS-GTR-42, vol. 1, 83 p. (Also available at http://www.fs.fed.us/rm/pubs/rmrs_gtr42_1.html).

Fire Effects on Flora

Brown, J. K., and J. K. Smith, eds. 2000. *Wildland Fire in Ecosystems: Effects of Fire on Flora.* USDA Forest Service General Technical Report RMRS-GTR-42, vol. 2. 257 p. (Also available at http://www.fs.fed.us/rm/pubs/rmrs_gtr42_2.html).

Fire Effects on Air

Sandberg, D. V., R. D. Ottmar, J. L. Peterson, and John Core. 2002. *Wildland Fire in Ecosystems: Effects of Fire on Air.* General Technical Report RMRSGTR-42-Vol. 5. Ogden, UT: U.S. Department of Agriculture, Forest Service, Rocky Mountain Research Station. 75 p. (Also available at http://www.fs.fed.us/rm/pubs/rmrs_gtr 42_5.html).

Other: Fire Effects on Aquatic Ecosystems

The journal *Forest Ecology and Management* devoted an entire issue in 2003 to present a compendium of fire effects on aquatic ecosystems—an area not covered in either the old or updated rainbow series.

Forest Ecology and Management (June 3, 2003) 178(1–2): 1–229. **"The Effect of Wildland Fire on Aquatic Ecosystems in the Western USA."** Edited by M. K. Young, R. E. Gresswell, and C. Luce.

Popular Magazines (Not Peer-Reviewed)

In contrast to scientific articles, contributions to popular magazines and mass media (such as newspapers, television coverage, and websites) may tend to editorialize or glamorize certain aspects of fire fighting or fire behavior with as much gloss as technical content. Some may be littered with biased views and unsubstantiated claims. In all cases, readers would do well to check data sources or to question assertions that are not supported by the materials presented. Similar cautions may pertain to websites. Informative articles may be found in the following:

American Forests, published by the American Forestry Association.

Fire Management Today: Earlier editions were released as *Fire Control Notes*, then *Fire Management Notes*, published by the USDA Forest Service, Washington Office, State and Private Forestry. In summer 2003, the magazine released the first of three volumes devoted to case studies of fire behavior. The entire three-volume set likely will become a collector's item for fire behavior specialists. Vol. 63(3), summer 2003, **"Fire Behavior Case Studies and Analyses: Part 1"** includes case study reprints of fires from 1937 to 1967. Vol. 63 (4), fall 2003, **"Wildland Fire Behavior Case Studies and Analyses: Part 2"** includes additional case studies, reporting standards, and advice for preparing future case study analyses. This volume also includes winning submissions from the 2003 photo contest. The third volume, Vol. 64(1), winter 2004, **"Forecasting Wildland Fire Behavior: Aids and Guides"** is devoted to knowledge-based protocols to assist analysts in forecasting fire behavior for safe and effective suppression strategies and tactics.

High Country News, published by the nonprofit High Country Association.

International Forest Fire News, published by the UN Food and Agriculture Organization/Economic Commission for Europe, Agriculture and Timber Division.

Sunset Magazine, published by the Sunset Publishing Corporation.

Wildfire, published by the International Association of Wildland Fire.

Other Selected References

Additional references for learning about fire include public service circulars, written articles, and educational tools such as fire games. Many videos, too numerous to summarize here, have been produced to describe fires and their management—including documentaries aired by commercial and public television stations, and educational footage made available by government agencies. These devices provide technical information (for example, on fire behavior prediction) and important tools for communicating with the public about safety and fire ecology. If discussed appropriately and in the right environment, wildland fire provides a rich context for general education. Fire education promotes understanding and integration of numerous concepts from many fields, including basic chemistry, physics, and biology, properties of matter and energy, ecosystem disturbances and cycles, plant and animal habitat and survival, and human interactions with natural ecosystems.

Fire and Public Safety

Davis, K. D., and R. W. Mutch. 1987. *Wildland Fire Hazards: Safety and Survival Guidelines for Recreationists and Homeowners.* USDA Forest Service, Fire Management Notes 48(2): 18–20.

U.S. Fire Administration. N.d. *Wildfire: Are You Prepared?* U.S. Fire Administration Publications Center (free pamphlet covering tips for protecting the home from wildfire, including creation of a safety zone, planning escape routes, and having a supply of necessities in the event of a wildfire). Available from http://usfa .fema.gov/applications/publications/.

Fire Behavior

Andrews, P. L. 1986. *BEHAVE: Fire Behavior Prediction and Fuel Modeling System—BURN Subsystem, Part 1.* USDA Forest Service General Technical Report INT-194, 130 p.

Byram, G. M. 1959. *"Combustion of Forest Fuels."* Pp. 61–89 in *Forest Fire: Control and Use,* edited by K. P. Davis. New York: McGraw-Hill Book Co., 584 p.

Finney, M. A. 1998. *Farsite: Fire Area Simulator—Model Development and Evaluation.* USDA Forest Service Research Paper RMRS-RP-4, 47 p.

Johnson, E. A., and K. Miyanishi, eds. 2001. *Forest Fires: Behavior and Ecological Effects.* London: Academic Press, 600 p.

Rothermel, R. C. 1991. *Predicting the Behavior and Size of Crown Fires in the Northern Rocky Mountains.* USDA Forest Service Research Paper INT-438, 46 p.

Scott, J. H. and E. D. Rhinehardt. 2001. *Assessing Crownfire Potential by Linking Models of Surface and Crown Fire Behavior.* USDA Forest Servicce Res. Pap. RMRS-RP-29, 59 p.

Fire Ecology

Bradstock, R. A., J. E. Williams, and M. A. Gill, eds. 2002. *Flammable Australia: The Fire Regimes and Biodiversity of a Continent.* Cambridge: Cambridge University Press, 472 p.

Johnson, E. A. 1992. *Fire and Vegetation Dynamics: Studies from the North American Boreal Forest.* Cambridge: Cambridge University Press, 129 p.

Tall Timbers Fire Ecology Conference Proceedings, Vols. 1–15 through 1974, then intermittently. Available (free, if still in print) from the Tall Timbers Research Station in Tallahassee, Florida (see Chapter 6 for contact information).

Walstad, J. D., S. R. Radosevich, and D. V. Sandberg. 1990. *Natural and Prescribed Fire in Pacific Northwest Forests.* Corvallis: Oregon State University Press, 317 p.

Fire Games

Smith, Jane Kapler, and Nancy E. McMurray. 2000. *Fireworks Curriculum Featuring Ponderosa, Lodgepole, and Whitebark Pine Forests.* General Technical Report RMRS-GTR-65. Fort Collins, CO: U.S. Department of Agriculture, Forest Service, Rocky Mountain Research Station, 270 p.

Note: FireWorks is an educational program for students in grades one through ten. The program consists of the curriculum in this report and a trunk of laboratory materials, specimens, and reference materials. It provides interactive, hands-on activities for studying fire ecology, fire behavior, and the influences of people on three fire-dependent forest types—*Pinus ponderosa* (ponderosa pine), *Pinus contorta var. latifolia* (interior lodgepole pine), and *Pinus albicaulis* (whitebark pine). The publication is downloadable at http://www.fs.fed.us/rm/pubs/rmrs_gtr65.pdf (approximately 4MB).

Living with Fire (http://www.fs.fed.us/rm/fire_game/).

Note: *Living with Fire* is an educational Internet game that puts you in the place of a fire manager, based on research and tools developed for real-world fire management. Recommended for ages ten and up. Last updated November 1, 2000.

Fire in the Tropics

Andersen, A. N., G. D. Cook, and R. J. Williams. 2003. *Fire in Tropical Savannas: The Kapalga Experiment.* New York: Springer-Verlag, 216 p.

Fire Weather

Schroeder, M. J., and C. C. Buck. 1970. *Fire Weather.* Agriculture Handbook 360. USDA Forest Service, 229 p.

Fuels Treatment

Graham, R. T., S. McCaffrey, T. B. Jain (technical editors). 2004. *Science Basis for Changing Forest Structure to Modify Wildfire Behavior and Severity.* General Technical Report RMRS-GTR-120. Fort Collins, CO: U.S. Department of Agriculture, Forest Service, Rocky Mountain Research Station. 43 p.

Pollet, J., and P. N. Omi. 2002. *Effect of Thinning and Prescribed Burning on Crown Fire Severity in Ponderosa Pine Forest.* International Journal of Wildland Fire 11(1): 1–10.

Omi, P. N., and E. J. Martinson. 2002. *Effects of Fuels Treatment on Wildfire Severity.* Final Report to the Joint Fire Science Program Governing Board. Western Forest Fire Research Center, Colorado State University, 40 p. Also available at http://www.colostate.edu./frws/research/westfire/FinalReport.pdf

History of Fire on Earth

Pyne, S. J. 1991. *Burning Bush: A Fire History of Australia.* 1991. Seattle: University of Washington Press.

———. 1995. *World Fire: The Culture of Fire on Earth.* New York: Henry Holt and Co.; paperback edition, University of Washington Press, 1997; Japanese edition, Hosei University Press, 2001.

———. 1997. *Vestal Fire: An Environmental History, Told through Fire, of Europe and Europe's Encounter with the World.* Seattle: University of Washington Press.

———. 2001. *Fire: A Brief History.* Seattle: University of Washington Press and British Museum, 2001.

———. 2003. *Smokechasing.* Tucson: University of Arizona Press.

Landscape Scale Studies Involving Consideration of Fire

Quigley, T. M., R. A. Gravenmier, S. J. Arbelbide, H. B. Cole, R. T. Graham, and R. W. Haynes. 1999. *The Interior Columbia Basin Ecosystem Management Project: Scientific Assessment.* CD-ROM available from:

Publications Distribution
Pacific Northwest Research Station
333 S.W. First Avenue
P.O. Box 3890
Portland, OR 97208–3890
503–808–2125

Sierra Nevada Ecosystem Project. 1996. By the Regents of the University of California (available at http://ceres.ca.gov/snep/).

Congressionally Commissioned Fire Studies

Government Accounting Office. 1999. *Western National Forests: A Cohesive Strategy Is Needed to Address Catastrophic Wildfires.* Report to the Subcommittee on Forests and Forest Health, Committee on Resources, House of Representatives. GAO Report 99–65, Washington DC, 64 p.

———. 2003. *Wildland Fire Management: Additional Actions Required to Better Identify and Prioritize Lands Needing Fuels Reduction.* Report to congressional requesters. GAO Report 03–805, Washington, DC, 67 p.

Nonprint Resources

Nonprint (tools and information) resources include models, websites, information clearinghouses, demonstration areas, and direct communications through workshops and conferences, or direct contacts with fire managers and scientists. Software and models available to fire managers are noted below, although some software requires advanced training and a few may not be accessible publicly. Much information is available on the worldwide web, accessible simply by typing the item of interest into a search engine. Useful representative websites are noted below. Governments, nongovernmental organizations (NGOs), and universities with fire academic/research programs also may maintain websites or extension programs for disseminating fire-related information. Public demonstration areas may be set up in local fuel treatment or recently burned areas to interpret and engage interest in fire and forest management. The Nature Conservancy has set up five demonstration areas: Long Island Pine Barrens, Jemez Mountains, Middle Niobrara-Nebraska Sandhills, Bighorns Landscape, and Upper Deschutes Basin.

In any given year, numerous workshops and conferences are devoted to fire management and fire ecology, covering regional, national, and international topics of interest. Geographic area fire council meetings provide important forums for managers, scientists, and academicians to meet and communicate new ideas. For example, the Interior West Fire Council convenes practitioners and researchers biannually from the U.S. and Canada. Some meetings are intentionally interdisciplinary, involving foresters, meteorologists, and fire ecologists, for example. Important information exchanges occur in formal presentations as well as informal conversations during break periods. Published proceedings add to the knowledge base, although considerable learning occurs by attendance and networking. Important nonprint resources, such as community partnerships, websites, conferences/workshops, and resource specialists; have become increasingly important to individuals and communities hoping to learn about wildland fire.

In general, websites produced and maintained by governmental entities are quite reliable and useful, but the proliferation of undocumented information on the web requires caution. The following are some of the best websites for keeping up with the latest information from a fire management standpoint, with emphasis on the United States. Several websites link to the same homepages, indicating the popularity and relevance of certain available reference sites. Two of the most useful websites for those getting started in fire studies are the National Interagency Fire Center (www.nifc.gov) and the USDA Forest Service Fire and Aviation Management (http://www.fs.fed.us/fire/) pages. Both pages contain useful information about national fire programs, insights into fire management and fire ecology, and numerous useful links. Other useful sites (arranged alphabetically) include the following:

Association for Fire Ecology (http://www.ice.ucdavis.edu/afe/)

The association is a national professional society for fire ecologists, providing an opportunity for managers, researchers, and educators to exchange ideas through national and regional conferences, an electronic newsletter and journal, and courses. Originally formed as the California Association for Fire Ecology, the association broadened its audience in 2000, with the first national congress on fire ecology. Student chapters promote the application of fire ecology through science and education at colleges and universities with active fire ecology programs.

California Department of Forestry and Fire Protection (http://www.fire.ca.gov/)

The men and women of the California Department of Forestry and Fire Protection (CDF) have fire protection and stewardship jurisdiction over 31 million acres of California's privately owned wildlands. In addition, the department provides varied emergency services in thirty-five of the state's fifty-eight counties via contracts with local governments. In a unique arrangement, CDF assumes protection responsibilities for some lower-elevation federal areas in exchange for the federal government's assuming lead responsibility on some of CDF's high-elevation properties. The department's firefighters, fire engines, and aircraft respond to an average 6,300 wildland fires each year. Those fires burn an average of nearly 144,000 acres annually.

Colorado Climate Center (http://climate.atmos.colostate.edu/)

The Colorado Climate Center assists the state of Colorado in monitoring climate over time scales of weeks to years, thereby contributing to a reduction in the state's vulnerability to climate variability and natural disasters, such as wildfires and floods. The center's web page provides links to regional climate centers and to the National Climate Data Center, where detailed climate information is provided. Other valuable climate links are listed as additional data resources.

Ecological Restoration Institute, Northern Arizona University (http://eri.nau.edu)

The Ecological Restoration Institute in the School of Forestry at Northern Arizona University supports ecological restoration activities such as education, research, and a forum for consideration of related issues. ERI develops comparative information on passive versus active management (thinning and prescribed fire), particularly in the dry ponderosa pine forest type of northern Arizona.

Fire and Fire Surrogate Research Study (http://www.fs.fed.us/ffs/)

This landmark study has established replicated treatment sites across the United States, all in low-severity fire regimes. Investigators are interested not only in the possible effects of silvicultural treatment (that is, mechanical thinning and prescribed fire) on wildfire severity but also in the treatment impact on such ecological attributes as nutrient cycling, seed scarification, plant diversity, disease and insect abundance, and wildlife habitat. Although still in progress, this study will add greatly to existing knowledge about the effects of treatments that attempt to mimic the ecological functions of low-intensity fires.

Fire Emissions Joint Forum (http://www.wrapair.org/forums/fejf/)

The Fire Emissions Joint Forum (FEJF) was formed to assist the Western Regional Air Partnership in addressing the Grand Canyon Visibility Transport Commission's (GCVTC) recommendations regarding wildfire, prescribed natural fire/wildland fire managed for resource benefits, prescribed fire, and agricultural fire. The forum addresses a broad definition of smoke effects that include consideration of public nuisance, public health, and visibility/regional haze.

Fire Effects Information System (http://www.fs.fed.us/database/feis/index.html)

The Fire Effects Information System (FEIS) provides up-to-date, online information about fire effects on plants and animals. It was developed at the USDA Forest Service Rocky Mountain Research Station's Fire Sciences Laboratory in Missoula, Montana. The FEIS database contains synoptic descriptions, taken from current English-language literature, of almost 900 plant species, about 100 animal species, and 16 Kuchler plant communities found on the North American continent. The emphasis of each synopsis is on fire and how it affects each species. Background information on taxonomy, distribution, basic biology, and ecology of each species is also included. Synopses are thoroughly documented, and each contains a complete bibliography.

Firebeaters (http://homepages.ed.ac.uk/ebfr89/firebeat/home.htm)

Firebeaters is a small group of ecologists, land managers, foresters, and conservationists interested in the history, ecology, control, and behavior of vegetation fires in the British Isles and North-West Europe. Objectives include to provide a forum for discussion of fire-related issues; to raise awareness of the importance of vegetation fires in land management and conservation; to promote the study of vegetation fires and help to identify research priorities; and to provide access to data and published information on the use and control of fire.

FireNet (http://sres.anu.edu.au/associated/fire/index.html)

An international fire information network established in 1993 and based at Charles Sturt University, Australia. Subscribers to FireNet are also able to participate in an electronic dialogue on topical and worldwide issues related to fires and their management.

Fire software downloads (http://fire.org)

The latest fire public access software is described and can be downloaded from the site maintained by Systems for Environmental Management (SEM), a private company with close ties to the Intermountain Fire Sciences laboratory in Missoula, Montana. SEM provides a valuable service to the fire management community by making the latest versions of fire software available for use and analysis, capitalizing on a strong working relationship with the Fire Sciences Laboratory where many of the computer programs are formulated and developed.

Firewise (http://www.firewise.org/)

The Firewise Home Page was created for people who live or vacation in fire-prone areas of North America; it is sponsored by the National Fire Protection Association (see Chapter 6), the American Red Cross, and the USDA Forest Service. Its intent is to allow residents and recreationists to make sensible choices about living in or using wild environments, otherwise known as the wildland/urban interface, urban interface, or exurban areas. Links from the home page include current information on regional workshops held annually throughout the United States.

Forest Fire Lookouts (http://www.firelookout.net/firelookout/firelookout.asp)

A complete listing is provided of nearly 500 lookouts in the United States and around the world, including listings on the National Historic Register, pictures, descriptions, map locations, and travel instructions. Partnership opportunities are described for lookout towers in need of maintenance and restoration. Link to Forest Fire Lookout Association website (http://www.firelookout.org/ffla/ffla.htm) to find out about rentals, job/volunteer listings, and general information on chapters in twenty-five states. Information on current operating lookouts may not be listed.

Four Corners Sustainable Forestry Partnership (http://www.fourcornersforests.org/)

In recent years, partnerships have proven essential to enabling communities to plan for eventual wildfire occurrence, to cope with recovery after a fire, to develop plans for defending communities and individual homes, for ensuring effective incident management during fires, and for implementing applied fire research. The Four Corners Partnership began in 1997 with the recognition throughout the Four Corners region (Colorado, New Mexico, Arizona, Utah) of increasing risks for catastrophic fire and insect outbreaks in forest ecosystems, as well as a declining capacity in communities to deal with forest restoration and maintenance needs. For another example of community partnerships, see the Quincy Library Group website noted below.

Global Fire Monitoring Center (http://www.fire.uni-freiburg.de/)

The GFMC provides the best information on global fire concerns, with important links to sources for international fire news. See additional descriptive information about the center in Chapter 6, including contact information.

Hayman Fire (2002) Case Study (http://www.fs.fed.us/rm/hayman_fire/)

In 2002, the western United States experienced an agonizingly long and protracted fire season because of unwanted ignitions following the prolonged drought. At the suggestion of Congressman Mark Udall (D-CO), the U.S. Forest Service commissioned a thorough review of causal factors that contributed to fire spread and effects. This report examines a range of issues raised by the Hayman fire, including the effectiveness of prefire fuel treatments, effects of weather, and socioeconomic impacts.

National Interagency Fire Center (http://www.nifc.gov)

The National Interagency Fire Center is the nation's nerve center when large fires are burning. It also maintains one of the most informative websites about fire, including a treasure trove of summary statistics, fire photos, fire situation updates, and related information. See http://www.nifc.gov/fireinfo/nfn.html for the latest information on ongoing fire incidents. Fire qualifications and firefighter safety are covered at http://www.nifc.gov/safety_study/quals-intro.html. Additional descriptive and contact information is included in Chapter 6.

National Oceanic and Atmospheric Administration (http://www.noaa.gov/)

The National Oceanic and Atmospheric Administration (NOAA) within the Department of Commerce provides numerous weather-related products of use to fire managers. For example, the Local Analysis and Prediction System (LAPS) integrates data from meteorological observation sites to produce gridded, three-dimensional profiles of weather phenomena that could alter fire behavior (http://laps.fsl.noaa.gov/cgi/laps_products.cgi). The Fire Consortia for Advanced Modeling of Meteorology and Smoke (FireCAMMs) relies on data and models from NOAA and other cooperators to produce fire weather analysis and forecasting tools (http://www.fs.fed.us/fcamms).

Operational Fire Maps (http://geomac.usgs.gov/#)

The Geospatial Multi-Agency Coordination Group, or GeoMAC, is an Internet-based mapping tool originally designed for fire managers to access online maps of current fire locations and perimeters in the conterminous forty-eight states and Alaska. Using a standard web browser, fire personnel can download this information to pinpoint the affected areas. With the growing concern over Western wildland fires in the summer of 2000, this application also became available to the public. GEOMAC is sponsored jointly by the U.S. departments of Interior and Agriculture (source: http://geomac.usgs.gov/AboutGeoMAC/WhatIs .html). Data layers include roads, cities, recent fire perimeters, thermal imagery (AVHRR, Modis), and nearby weather stations. In order to give fire managers near real-time information, fire perimeter data is updated daily based upon input from incident intelligence sources, GPS data, and infrared (IR) imagery from fixed-wing and satellite platforms.

Program for Climate, Ecosystem, and Fire Applications (http://www.cefa.dri.edu/)

CEFA is an abbreviation for the Program for Climate, Ecosystem, and Fire Applications. The program was formed on October 1, 1998, through an assistance agreement between the Bureau of Land Management Nevada State Office and the Desert Research Institute (DRI). As of November 2000, a new five-year assistance agreement was signed with the BLM national Office of Fire and Aviation to continue basic climate studies and product development for fire management at the national level. CEFA resides within the Division of Atmospheric Sciences of DRI and works closely with the Western Regional Climate Center (WRCC).

Quincy Library Group (http://www.qlg.org/pub/contents /overview.htm).

Founded in 1992 by a timber industry forester, a county supervisor, and an environmental attorney, this private-public coalition addresses a variety of common environmental concerns (for example, the California spotted owl, water, logging, fire and fuels management) in northeastern California, encompassing Lassen, Sierra, and Plumas counties. Since then the QLG's area of interest has grown to encompass parts of eight counties and three national forests with funding provided by its inclusion in the FY1999 Omnibus Spending Bill by the U.S. Congress. For another example of private and public partnerships, see the Four Corners Sustainable Forestry Partnership website (see page 306–307).

Smokey Bear (http://www.smokeybear.org, also http://www .smokeybear.com)

In 1950 a bear cub was rescued from a charred tree within a wildfire in New Mexico. Originally dubbed "Hotfoot Teddy" by his caretakers, he soon was renamed "Smokey Bear," after a poster produced in 1945 by Albert Staehle (originated and authorized by the USDA Forest Service and the Advertising Council). Smokey Bear has become a national symbol of fire prevention, recognized worldwide as an important tool for public education.

Synoptic Weather Patterns Associated with Critical Fire Weather (http://climate.usfs.msu.edu/climatology)

Atmospheric conditions play a critical role in affecting the behavior and severity of wildland fires, as well as the probability of their occurrence. Fire managers rely heavily on current observations and forecasts of local, regional, and synoptic atmospheric conditions to prepare for and carry out fire suppression and prescribed fire activities. Synoptic weather maps are critical for understanding the relationships between large-scale upper and middle atmospheric processes to regional-scale fire-weather systems. These maps help establish reliable forecasts of fire-weather and for understanding how potential large-scale changes resulting from a globally changed climate might influence fire occurrence and severity in the United States.

USDA Forest Service Fire and Aviation Management (http://www.fs.fed.us/fire)

The USDA Forest Service Fire and Aviation Management website provides recent fire information, forest management, and a forests/people page (including links to wildland fire assessment, fire effects information, tools, and numerous other useful sites).

Interesting national coarse-scale maps of fire regimes and fuel conditions are available at http://www.fs.fed.us/fire/fuelman/. The Wildland Fire Assessment System (www.fs.fed.us/land /wfas/) provides national fire danger rating maps, including "greenness," available from satellite imagery. Additional information on fire systems and tools, fire news and publications, fire centers and research, and links is available at http://www .fs.fed.us/land/#fire (accessed January 2, 2002).

U.S. Large Fire Maps (http://activefiremaps.fs.fed.us/)

National maps of large fires (that is, fires larger than 10,000 acres) are provided from the USFS Remote Sensing Application Center, using the Moderate Imaging Spectralradiometer (MODIS) sensors aboard NASA's Earth Observing System satellites (Terra and Aqua). The MODIS instrument maps the entire surface of the earth once every other day (daily in the Northern Hemisphere) at moderate spatial resolutions (250, 500, and 1000 m). See also http://activefiremaps.fs.fed.us/fire_imagery.php, for MODIS active fire imagery.

Western Forest Fire Research Center (WESTFIRE), Colorado State University (http://www.cnr.colostate.edu/frws/research /westfire/index.htm)

Colorado State University has one of the largest university programs (undergraduate and graduate teaching plus active research programs) in forest fire science in the nation, if not the world. WESTFIRE provides an organizational umbrella to the numerous fire-related research and instructional programs available at the university.

Western Regional Climate Center (http://www.wrcc.dri.edu/)

The Western Regional Climate Center (WRCC) is a repository for data and information on climate in the western United States. WRCC programs provide information for better understanding of fire weather and climatic variations in the West, including, for example, past El Niño and La Niña occurrences that can be linked to wildfire outbreaks.

The Wildfire Automated Biomass Burning Algorithm (http: //cimss.ssec.wisc.edu/goes/burn/wfabba.html)

This is an experimental program to spot and map fires in the Western Hemisphere using satellites located more than 20,000 miles above the earth. The geostationary NOAA weather satellite GOES-12 provides coverage for North and South America, while GOES-10 covers North America only. Links from the homepage lead to the most recent image and provides the option to go to higher-resolution images of different regions, as well as loops of the images covering the last several hours (or days, depending on the region).

Wildfire News (http://www.wildfirenews.com/)

The Wildfire News website includes recent news and articles of interest to laypersons, as well as pages primarily for wildland firefighters, including hot-shots and smokejumpers. This page is also a good source for fire photos, job listings, current controversies, and links to other information.

Note: Other useful websites are included with agency and organization contact information in Chapter 6.

Software

The software described below is generally public-accessible and available by consulting FRAMES, the home for Fire Management Tools Online (http://www.frames.gov/tools/). Software programs are generally designed for use on an IBM PC-compatible computer. To take full advantage of the interactive interface, the PC must be running Windows 95/98 or Windows NT 4.0, or higher. The use of the interactive interface also requires an Internet web browser such as Microsoft Internet Explorer or Netscape Navigator (version 3.04 or greater for both browsers—version 4.0 or greater is recommended). No recommendations are intended regarding compatibility issues, nor is endorsement implied for private party or government software.

BAER. Last updated: February 16, 2000.

Research publication synthesizing thirty years of USDA Forest Service Burn Area Emergency Rehabilitation (BAER) projects.

BEHAVE. Last updated: August 21, 2003.

The BEHAVE Fire Behavior Prediction and Fuel Modeling System gathers available fire models into a system that is driven by direct user input. The fire modeling capabilities have been significantly expanded and the user interface vastly improved with the updated BehavePlus Fire Modeling System.

CONSUME. Last updated: October 27, 2003.

Consume 2.1 is a PC-based, interactive fuel consumption model that predicts total and smoldering fuel/biomass consumption during prescribed fires and wildland fires. Predictions are based on weather data, the amount and fuel moisture of fuels, and a number of other factors, with primary focus on the Pacific Northwest.

CPS. Last updated: September 01, 2001.

The focus of the Campbell Prediction System (CPS) is to predict changes in fire behavior so that firefighters can get out of harm's way before the fire makes its move. The book has maps that display the trigger points, tracks, and time tags that aid in tactic selection. Shaded terrain maps and solid terrain models make for better understanding of the terrain and why some tactics work and others don't.

EPM. Last updated: February 14, 2000.

The Emissions Production Model (EPM) predicts air pollutant emissions source strength, heat release rate, and plume buoyancy consistently for all fire environments and fuel types. It requires an estimate of flaming and smoldering consumption, and a stylized description of ignition pattern. EPM then calculates timed emission rates for gases, particles, and heat.

FARSITE. Last updated: February 14, 2000.

FARSITE is a fire growth simulation model. It uses spatial information on topography and fuels along with weather and wind files. FARSITE incorporates the existing models for surface fire, crown fire, spotting, and fire acceleration into a two-dimensional fire growth model. FARSITE runs under Microsoft Windows operating systems (Windows 3.1x, 95, NT) and features a graphical interface. Users must have the support of a geographic information system (GIS) to use FARSITE, because it requires spatial landscape information.

FDRPC. Last updated: March 05, 2002.

Fire Danger Rating Pocket Card for Firefighter Safety (FDRPC) allows local or regional fire managers to prepare cards that describe historical fire weather and incidents for distribution to firefighters, especially those who may be unfamiliar with local conditions.

FEIS. Last updated: May 14, 2003.

The Fire Effects Information System (FEIS) is a computerized encyclopedia of information describing the fire ecology of more than 1,000 plant and animal species and plant communities. FEIS summarizes current information about fire effects on plants and animals.

FEMHB: Fire Effects Monitoring Handbook (available on the NPS Wildland Fire Analysis Software CD, 1999).

FFE-FVS. Last updated: June 14, 2002.

The Fire and Fuels Extension to the Forest Vegetation Simulator (FVS) incorporates models of fuel dynamics, fire behavior, and fire effects into a base model of forest stand development. Effects of timber harvest, fuel treatment, and fire on subsequent fuel dynamics, stand development, and potential fire intensity can be simulated and displayed graphically for a period of decades.

FireAway. Last updated: November 13, 2000.

A subset of the Behave software, plus some additional tools and calculations from the Fireline Handbook, that runs on a handheld Palm computer.

FireDirect. Last updated: April 15, 2002.

FireDirect from RedZone Software is mapping software providing innovative GIS capabilities for agencies fighting fires in the Wildland/ Urban Interface. See the website at www.redzonesoftware .com.

Firefamily+. Last updated: December 02, 2002.

Provides historical analysis programs that operate against the fire weather and fire occurrence databases in the National Integrated Fire Management Interagency Database (NIFMID). The programs whose functionality are incorporated (and expanded) in Firefamily Plus include PCFIRDAT: A non-Y2K compliant PC-DOS program that processes historical fire weather data. It can produce daily lists, frequency distributions of weather, and fire danger indexes. It also produces a "passing file," which is a flat file of the daily list information in a fixed field format. PCSEASON: A non-Y2K compliant PC-DOS program that reads passing files and does primitive graphics, conditional climatology, seasonal severity, and RERAP reports. FIRES: A non-Y2K compliant PC-DOS program that reads passing files and fire occurrence files and does analysis of index or weather variable performance in terms of the relationships between seasonal index traces and fire occurrence potential. CLIMATOLOGY: A set of non-Y2K compliant PC-DOS programs that provide detailed climate information by ten-day and monthly periods.

fireLib. Last updated: April 12, 1999.

fireLib is a C function library for predicting the spread rate and intensity of free-burning wildfires. While fireLib is a direct descendant of the BEHAVE fire behavior algorithms for predicting fire spread, it is optimized for highly iterative applications such as cell- or wave-based fire growth simulation.

FIRES. Last updated: April 12, 1999.

Fire Information Retrieval and Evaluation System (FIRES) provides methods for evaluating the performance of fire danger rating indexes. The relationship between fire danger indexes and historical fire occurrence and size is examined through logistic regression and percentiles. Historical seasonal trends of fire danger and fire occurrence can be plotted and compared. Methods for defining critical levels of fire danger are provided.

FireTower. Last updated: July 11, 2000.

An Interactive Software Tool for Simulating the Movement of Fire Through Landscapes.

FIREWORKS. Last updated: March 11, 2001.

A portable trunk that contains educational materials for hands-on learning about how forests change over time, especially in relationship to fire. Provides curricula for all grade levels. For use by classroom teachers and agency staff.

FMAPlus. Last updated: February 24, 2003.

A software package designed to make life a lot easier for calculating planar intercept and photo series inventories (DDWoodyPC), using digital photo series (PSExplorer), and calculating canopy weights, canopy base height, and canopy bulk density (CrownMass) for fire behavior predictions. The fuel model manager program (FMMgr) allows the user to create custom fuel models to fit the local area.

FOFEM. Last updated: May 24, 2002.

First Order Fire Effects Model (FOFEM) is an easy-to-use computer program for predicting effects of prescribed fire and wildfire. FOFEM predicts fuel consumption, smoke production, and tree mortality. Area of applicability is nationwide on forest and nonforest vegetation types. FOFEM also contains a planning mode for prescription development.

I-SUITE. Last updated: September 9, 2002.

A group of applications for automating incident operation, using a centralized database of resources assigned to the incident.

IIAA. Last updated: April 12, 1999.

Interagency Initial Attack Assessment (IIAA) is a tool used to develop budget requests as part of the National Fire Management Analysis System (NFMAS) process.

KCFAST. Last updated: April 12, 1999.

KCFast is a menu-based computer application that simplifies data retrieval from the National Interagency Fire Management Integrated Database (NIFMID) for fire weather and fire occurrence.

MfFSF. Last updated: July 20, 2001.

Meteorology for Fire Severity Forecasting includes monthly fire weather forecasts that can provide fire managers with a quick and easy planning tool. It is based on the monthly forecast of 700 millibar heights issued by the NWS Climate Analysis Center in Washington, DC.

NEXUS. Last updated: March 26, 2001.

NEXUS is an EXCEL spreadsheet that links surface and crown fire prediction models for assessing fire behavior, including possible links to stand structural characteristics such as crown base height.

NFSPUFF. Last updated: April 12, 1999.

NFSPUFF is a screening/planning level, three-dimensional, gridded wind field smoke emissions and trajectories puff model. It is designed to predict ground level concentrations of particulate matter and gaseous pollutants from multiple sources in complex terrain in the western United States. The model incorporates an emission production module (EPM) with National Weather Service predictions for upper-air winds, extrapolated to the surface, to predict potential pollutant transport. Tabular, 2-D and 3-D graphics are displayed.

PCDANGER. Last updated: April 12, 1999.

PCDANGER is a Personal Computer (PC) application of the National Fire Danger Rating System (NFDRS) that calculates both 1978 and 1988 version fire danger indexes from daily weather observations and forecasts.

PCHA. Last updated: April 12, 1999.

Personal Computer Historical Analysis (PCHA) is a personal computer (PC) program developed to complete the Historical Analysis required for the National Fire Management Analysis System (NFMAS). PCHA is a tool designed to help analyze historical wildland fire occurrence for fire planning purposes.

PLUMP. Last updated: February 14, 2000.

Plump calculates plume rise from large fires, including pyrocumulus and cumulus growth without fire, based on a unidimensional time-dependent model including principles of cloud physics and entrainment.

PWA99 Prevention Workload Analysis (available on the NPS Wildland Fire Analysis Software CD 1999).

RAfFS. Last updated: February 14, 2000.

Articles and information discussing recent fire research and its application to firefighter safety.

RERAP. Last updated: January 27, 2000.

Rare Event Risk Assessment Process (RERAP) is a Windows-based program that helps calculate the information needed to manage prescribed fire and wildfires, based on projections of long-term fire spread and the occurrence of a season-ending precipitation event.

SASEM. Last updated: December 3, 2002.

SASEM is a Gaussian dispersion model designed to predict ground level particulate matter and visibility impacts from single sources in relative flat terrain in the western United States.

SIAM. Last updated: November 10, 1999.

Structure Ignition Assessment Model. SIAM assesses potential residential ignitions during wildland/urban interface (WUI) fires given a structure's materials and design and its exposure to flames and firebrands to produce an index of WUI ignition risk.

TAC-PAK. Last updated: December 31, 2002.

Emergency Operations Center in a Briefcase. Complete incident command and worldwide communications from a battery-powered briefcase package

VALBOX. Last updated: April 12, 1999.

Ventilated Valley Box Model (VALBOX) is a screening model designed to predict ground level concentrations of particulate matter and gaseous pollutants under stagnation conditions in mountain valleys.

Ventura_Tools. Last updated: May 20, 2002.

A complete set of ArcView 3.2 extensions to create and edit critical incident data. After the data have been developed, a quick map layout tool is employed to create IAP, Planning and Public Display Maps.

WFAS. Last updated: April 10, 2000.

WFAS: Wildland Fire Assessment System is an Internet-based system that provides national fire potential and weather maps.

WFSA_Plus03. Last updated: August 7, 2003.

WFSA Plus03 is a decision support software package designed to assist managers in developing strategies for a particular fire incident, including documentation for the Wildland Fire Situation Analysis and Wildland Fire Use assessments and plans.

Glossary

Active crown fire   A crown fire whose spread is linked to surface fire intensity.

Air mass   An extensive body of air having the same properties of temperature and moisture.

Airtanker   Planes fitted with tanks for transporting fire retardant or water to a fire. Capacities range from 2,000 to 3,000 gallons per aircraft.

Anchor point   A secure starting point for building and holding a fireline.

Area ignition   Simultaneous ignition of a large area because of accumulation of combustible gases or wind shifts.

Atmosphere   The envelope of air surrounding the earth.

Atmospheric stability   The propensity of the atmosphere to move vertically in response to a lifting (or lowering) force. An unstable environment will augment formation of a well-formed convection column above a fire and support extreme burning conditions. A stable environment may actually constrain burning activity.

Available fuel   The amount of fuel that actually burns in a fire, subject to burning and fuel conditions.

Backfire   An aggressive suppression tactic usually employed as a last resort to stop, delay, or split an oncoming wildfire front. Unburned fuel between the oncoming fire and secured fire line is ignited precisely at the time when the backfire will be drawn by the convection column of the wildfire, thereby removing fuel and changing spread direction and effects. Backfiring is executed as a command decision.

Backing fire   A fire, or that part of a fire, that is spreading against the wind or down a slope.

Bladder bag   A portable rubberized bag of water (5 gal.), equipped with hose and nozzle and fashioned with backpack straps, for use in the mopup stages of a fire.

Blowup   A sudden change in fire intensity or rate of spread that can catch firefighters unaware.

Brush hook   A cutting tool with a hooked cutting edge, adapted primarily for cutting brush or chaparral.

Burning out   Fires intentionally set to remove fuels inside the control line, usually done by the crew boss.

Burning period   That part of each 24-hour day when fires will spread most rapidly, typically from about midmorning to sundown.

Cambium   The active growth tissue just beneath the tree bark.

Canopy   The stratum of fuels containing the crowns of the tallest vegetation.

Cellulose   Glucose units that constitute the chief part of plant cell walls and provide structure to vegetative materials, such as forest fuels.

Chinook, or Chinook wind   A foehn or gravity wind, blowing down the eastern slopes of the Rocky Mountains.

Combustion   The rapid oxidation of material to produce heat energy.

Conduction   Heat transfer through a solid—i.e., by molecular activity.

Cone serotiny   An attribute of trees, such as lodgepole pine, whereby the cones remain closed until opened by heat.

Contained fire   A fire that is surrounded by constructed fire lines and natural barriers.

Convection   Heat transfer through a gas or liquid, resulting from vertical motion.

Convection column   The thermally-produced, billowing, vertical column of smoke, ash, and debris that forms above a fire.

Crown fire   A fire that involves the live crown of trees or shrubs, spreading through the forest canopy more or less independently of the surface.

Dendrochronology   Methods for dating tree rings for cross-dating with fire scars.

Direct attack   A suppression method that involves work along the burning edge, using wetting, cooling, smothering, chemical quenching, or mechanical separation of unburned fuels.

Duff   A mat of partially decomposed organic matter from fallen foliage, herbaceous vegetation, and decaying wood, located immediately above the mineral soil.

Equilibrium moisture content　A threshold of moisture wherein dead fuels neither gain nor lose additional moisture with time, under constant atmospheric pressure, relative humidity, and temperature.

Ether extractives　Volatile chemicals (soluble in ether solutions) that enhance combustion; extractives usually impart an oiliness (or terpene fragrance) that can be detected in the field.

Extended attack　Strategy and tactics employed on a fire that is not contained or controlled by initial attack forces.

Fire behavior　The manner in which a fire reacts to fuel, air mass, and topography, resulting in spread, intensity, and other fire characteristics.

Firebrand　An ember or particle of fuel ignited by an intense heat source that can ignite unburned fuel, often emitted ahead of the fire front and causing **spot fires**.

Firebreak　A preconstructed or natural break in fuel continuity, initially cleared and maintained to bare mineral soil.

Fire characteristics chart　A two-dimensional graphic for rating fire behavior and suppression difficulty, based on rate of spread and heat per unit area (or **fire intensity**).

Fire containment　Encircling a fire perimeter with **fireline** constructed by firefighters.

Fire danger　Potential for an initiating fire to cause management problems.

Fire danger rating　A quantitative assessment of fire potential, mostly related to weather and climatic influences over a broad area.

Fire entrapment　A situation in which personnel are unexpectedly caught in a fire behavior–related, life-threatening position where planned escape routes or safety zones are absent, inadequate, or compromised. An entrapment may or may not include deployment of a fire shelter. These situations may or may not result in injury; they include "near misses."

Fire environment　The surrounding conditions (fuel, air mass, topography) that influence the behavior of fires.

Fire intensity　Rate of energy release from a fire's flaming front, useful for gauging the suppression difficulty and above-ground fire effects.

Fireline or fire line　A barrier to fire growth constructed by firefighters using tools to remove flammable fuels.

Fireline intensity　The rate of energy release during combustion per unit length of fire front; also called Byram's intensity.

Fire regime　A concept for understanding the role of fire in an ecosystem, including descriptors for historical fire frequency, intensity, size, and seasonality.

Fire return interval   The average number of years elapsing between recurrent fires in an area.

Fire season   The period(s) of the year during which wildfires are likely to occur, spread, and produce sufficient damage to warrant organized fire control.

Fire severity An indicator of the ecological impact of a fire or set of fires, indicating effects on vegetation or degree of fuel consumption, or integrating above-ground and below-ground heat effects.

Fire shelter An aluminized pup-tent offering protection by means of reflecting radiant heat and providing a volume of breathable air in a fire entrapment situation. Fire shelters should be used only in life-threatening situations, as a last resort.

Fire storm Violent convection caused by a large continuous area of intense fire. Often characterized by destructively violent surface indrafts, near and beyond the perimeter, and sometimes by tornadolike whirls.

Fire triangle   An instructional device using the legs of a triangle to depict the role of oxygen, heat, and fuel in combustion.

Flame height The height above ground of the flame tip (not to be confused with flame length).

Flame length The distance measured from the tip of the flame to the middle of the flaming zone at the flame base; typically measured in meters or feet.

Flaming front The zone of a moving fire within which the primary combustion mechanism is active flaming. Behind this zone, combustion is primarily smoldering and glowing.

Flank fire   A fire set parallel to the wind and allowed to spread at right angles to control lines.

Flanks of a fire Portions of the fire perimeter that are roughly parallel to the main direction of fire spread; also called the sides of the fire.

Fuel   Live and dead vegetation that will burn, including coarse woody debris, understory vegetation, tall shrubs, tree crowns, duff, and litter.

Fuel availability   See **available fuel**.

Fuelbreak   A preconstructed area in which fuels have been converted to lower volume, less flammable fuel types in order to facilitate fire control or prescribed burning operations.

Fuel moisture   The amount of water in a fuel particle, expressed as a percentage of the particle's oven-dried weight.

Fuel profile   Describes the ground, surface, and crown attributes that support combustion.

Hazard Properties of a fuels complex related to ignition wildfire behavior and severity, or suppression difficulty.

Head fire A fire spreading with the wind or up a slope.

Heat transfer Mechanisms by which heat is transferred: typically radiation, convection, and conduction.

Helibase The main location within the general incident area for parking, fueling, maintaining, and loading helicopters. The helibase is usually located at or near the incident base.

Helispot A temporary landing spot for helicopters, often natural openings or constructed by helitack crews.

Helitack, Helitack crew Specially trained firefighters who are transported to fires via helicopters. Crew members are familiar with the tactical and logistical advantages of using a helicopter to transport personnel and equipment to an ongoing incident. Crewmembers may be trained to rappel from a hovering helicopter in terrain that does not permit a landing.

Holdover fire Fire that lies dormant for a considerable time after ignition (also called a **sleeper fire**).

Hotshot crew Highly trained and experienced crews of firefighters (usually twenty persons) typically given the more challenging assignments on fires, including fireline construction, firing operations, and mopping up. Crew members usually use a wide variety of specialized tools, including shovels, McLeods, pulaskis, brush hooks, chainsaws, and possibly fireline explosives.

Hotspotting A firefighter's tactic for cooling down a fire by using water or digging and throwing soil at the base of a flame in one smooth motion to rapidly remove available fuel feeding the flame.

Hygroscopic Property of dead fuels that result in moisture exchange (increase, decrease) with changes in environmental moisture.

Incident action plan A document containing general suppression objectives that reflect overall strategy for an incident; details specific suppression plans for the next operational period.

Incident command system (ICS) The combination of facilities, equipment, personnel, procedure, and communications operating within a common organizational structure, with responsibility for the management of assigned resources to effectively accomplish stated objectives pertaining to an incident.

Independent crown fire A crown fire that spread indepently of surface fire intensity.

Indirect attack A method of suppression in which the control line is established mostly along natural firebreaks, water bodies, or at considerable distance from the fire, and all intervening fuel is backfired or burned out.

Initial attack   Activities undertaken by the first suppression resources that arrive at a fire.

Inversion, in the atmosphere   Warm air lying above cold air, a situation often encountered above mountainous terrain during early morning hours.

Ladder fuels   Continuous flammable fuels (e.g., branches, needles, lichen, and the like) linking surface and crown strata. The presence of fuel ladders during peak burning conditions increases the likelihood of crown fire initiation and spread.

Line ignition   Simultaneous ignition by a continuous front of fire, as opposed to a single point.

Mass fire   A large fire or fires burning with high intensity.

McLeod   A combination tool with heavy-duty rake and sharpened hoe.

Megafire   Large project fires that require more people, more equipment, and greater commitment of financial resources.

Mopup   A latter stage of fire suppression operations in which fuels are not actively burning and managers consider the fire to be approaching controlled status or extinction.

MREs   Meals ready to eat, a staple of firefighters when centralized food preparation facilities have not been established on a fire.

Nomex   Heat-resistant material used to fabricate firefighter protective clothing (e.g., shirts, pants, gloves, hood).

Overhead   A team that administers an incident.

Passive crown fire   A fire that torches in individual trees. (See **active crown fire** and **independent crown fire**).

Plume dominated fire   A fire burning with well-formed convection column, spreading in response to updrafts and downdrafts created by buoyancy in an unstable atmosphere, possibly including collapses of the column caused by gravity. Contrast with a wind-dominated fire.

Point ignition   Ignition of an individual point; the result of human or natural causes, as compared with **line ignition.**

Prescribed fire The intentional use of fire to achieve specific objective(s). Fire behavior and desired size and effects of treatment are controlled through ignitions under prespecified fuel, weather, and topographic conditions (fuel moisture, windspeed), along with appropriate ignition techniques.

Prescription window   Specified conditions under which a prescribed fire is ignited.

Probability of ignition   The likelihood that a firebrand (glowing or flaming) will cause a fire once it encounters receptive fuels; usually calculated from temperature, relative humidity, and fine fuel moisture content.

Project fire   An extended attack fire that requires a more complex organization including a fire camp (also called a *campaign fire*).

Pulaski   A combination tool (ax and mattock, or grubbing head).

Pyrolysis   The process by which solid fuel is preheated and decomposed to volatile gases as a precursor to ignition and flaming.

Radiation   Heat transfer by electromagnetic waves moving at the speed of light, such as from the sun to the earth's atmosphere.

Radiosonde   A meteorologist's device carried aloft and by a balloon equipped with measuring devices for transmitting temperature, pressure, and humidity derivative information to a ground recorder.

Red card   Certification of firefighter qualifications to perform specific tasks during a wildfire. Training and prior experience requirements vary by position, being lowest for basic firefighting and higher for skill positions within a large fire organization.

Red flag warning   A term used by forecasters to draw attention to weather conditions of limited duration that may result in extreme burning conditions.

Relative humidity   A measure of the moisture or water vapor pressure in the atmosphere compared with conditions at saturation.

Retardant   Chemical solution used in fire suppression activities, typically diammonimum phosphate (a soil fertilizer) colored with a biodegradable marker dye (pink) when dropped from an aircraft. Chemical retardants are occasionally used in ground-based fire engines.

Risk   A causative agent, such as lightning or humans; also refers to the likelihood of a fire ignition in an area.

Savanna   Grassland with intermittent trees and shrubs.

Shovel   A combination tool with a cutting edge (for chopping down small trees, and cutting limbs and roots); also used for scraping/digging.

Slash, logging slash, thinning slash   Residual branches, twigs, needles, and other fuels from a logging or thinning activity.

Smokejumpers   Airborne firefighters who parachute from planes to attack fires in remote or otherwise inaccessible areas.

Spike camp   A makeshift camp for temporary lodging and feeding of firefighters, prior to establishment of a fire camp.

Spot fire   A fire that starts apart from the main body, for example by windborne embers.

Thermal belt   A local region in mountainous terrain that typically has the least variation in diurnal temperatures, the highest nighttime temperatures, and the lowest relative humidities.

Timelag   An indication of the rate at which dead fuel gains or loses moisture because of environmental changes. The time (in hours) necessary for a fuel particle to gain or lose approximately 63 percent of the difference between its initial content and the equilibrium moisture content.

Torching   Intermittent ignition of tree and shrub crowns from an advancing surface fire.

Wildfire   An unplanned wildland fire requiring suppression action, one that may or may not be affected by suppression activities.

Wildland fire use, Wildland fire managed (used) for resource benefit Lightning fires that are allowed to burn in order to achieve resource benefit; may require herding or partial suppression.

Wildland fire situation analysis A required document describing the complexity of an incident, management alternatives, and a selected course of action for managing an incident.

Wind dominated fire   A fire whose spread is dominated by the wind, as opposed to a **plume dominated fire.**

Index

About the Author

Philip N. Omi is professor of forest fire science and director of the Western Forest Fire Research Center at the Colorado State University, Fort Collins, Colorado. Dr. Omi has taught undergraduate and graduate courses related to wildland fires since 1977, while overseeing one of the largest instructional and research academic programs in the United States. During 1987–1988, he served as visiting professor of wildland fire management at the University of Washington, Seattle. He has been studying forest fires for thirty years, including forty-three research projects covering a variety of topics related to fire science and management. Prior to earning his Ph.D. (in wildland resource science) at the University of California, Berkeley, he worked as a seasonal firefighter in northern and southern California for four years, including assignments throughout the western United States. He has authored or coauthored more than 100 journal articles and technical papers related to fire science during his career.